MICROBIOLOGY RESEARCH ADVANCES

PATHOGENIC BACTERIA

PATHOGENESIS, VIRULENCE FACTORS AND ANTIBACTERIAL TREATMENT STRATEGIES

MICROBIOLOGY RESEARCH ADVANCES

Additional books and e-books in this series can be found on Nova's website under the Series tab.

MICROBIOLOGY RESEARCH ADVANCES

PATHOGENIC BACTERIA

PATHOGENESIS, VIRULENCE FACTORS AND ANTIBACTERIAL TREATMENT STRATEGIES

KEITH D. WATTS
EDITOR

nova
Medicine & Health
New York

DOI: https://doi.org/10.52305/BNPS2149

NOTICE TO THE READER

Additional color graphics may be available in the e-book version of this book.

Library of Congress Cataloging-in-Publication Data

ISBN: 978-1-68507-422-7

Published by Nova Science Publishers, Inc. † New York

Contents

PREFACE

Although most forms of bacteria are harmless, pathogenic bacteria can cause disease and as such these microorganisms warrant scientific attention. Chapter one describes the virulence problem caused by extracellular polymeric substances and antibacterial compounds capable of inhibiting their production, biofilm formation, and bacterial pathogenicity. Chapter two provides an overview of biophysical techniques that can validate targets of phytochemicals with anti-virulence properties that can help elucidate their mode of action. Chapter three discusses terpenes' anti-virulence modes of action, inhibition of biofilm, disruption of the cell wall and bacterial membrane, intercellular leakage, interruption of nucleic acids transcription, and interference with the quorum sensing signaling process. Chapter dour investigates the anti-pathogenic potential of essential oil from the plant asafoetida obtained through microwave-assisted extraction method. Finally, chapter five deals with the effect of audible sound waves on bacteria.

Chapter 1 - The persistence and resistance of bacteria to disinfection processes are associated with their ability to form biofilms on biotic and abiotic surfaces. Bacteria adhere to natural and engineered surfaces and become mature biofilms encased in self-produced extracellular polymeric substances (EPS). EPS consists of polysaccharides, proteins, metabolites, and nucleic acids. It is well known that cells in biofilms are significantly more resistant to disinfectants and external attacks. It should be noted that EPS may also be involved in cell defense. One of the main functions of EPS is to prevent the penetration of dangerous molecules into bacterial cells, interacting with chemicals making them non-bioavailable, and degrading/hydrolyzing dangerous compounds. This characteristic makes EPS potentially dangerous in the field of food and medicine. Exopolysaccharides are formed in the biofilm of many bacterial species such as *Escherichia* spp., *Salmonella* spp., *Bacillus* spp., *Pseudomonas* spp., *Klebsiella* spp., and *Enterobacter* spp. In the case of *Salmonella*, cellulose is one of the main polymeric substances produced by the cellulose synthase enzymatic complex, made up of glycosyltransferases. Therefore, a mechanism for controlling biofilms can be inhibiting this enzyme, and consequently, reducing pathogenicity. The EPS of *P. aeruginosa* predominantly consists of different polysaccharides and proteins. Mucoid strains of *Pseudomonas* spp. are characterized by the overproduction of alginate, an extracellular polysaccharide, increasing its pathogenicity. Alginate is an important virulence factor in chronic lung infections in patients with the inherited disease cystic fibrosis, and it is the main reason for cystic fibrosis patients' premature death. Compounds obtained from natural sources, enzymatic treatments, chemical synthetic, and nanoparticles have shown effectiveness in inhibiting EPS of pathogenic bacteria. Several essential oils can inhibit the formation of biofilms, and this effect was related to the inhibition of the EPS production mechanism. In general, EPS as mediators of biofilm formation and stability should be considered target structures for corrective actions to eliminate, prevent, or control undesirable biofilms and bacteria-influenced corrosion of materials, both in the food industry and in the field of medicine. Therefore, the objective of this chapter is to describe the virulence problem caused by EPS and antibacterial compounds capable of inhibiting their production, biofilm formation, and bacterial pathogenicity.

Chapter 2 - Antibiotic resistance is one of the major threats to global health; therefore, the challenge is developing strategies to attack this problem. Anti-virulence therapy is an alternative that has gained attention in the last years, and it is aimed to attenuate the bacterial strategies to infect and cause disease, and sometimes it does not aim to affect the pathogen viability. Phytochemicals have been demonstrated anti-virulence properties against a wide range of pathogenic bacteria. Studies on this topic are commonly focused on the characterization of physiological changes of treated bacteria such as bacterial membrane damage, biofilm inhibition or changes in expression of virulence-related genes. However, few studies delve into the molecular targets of these compounds and their molecular interactions. In this sense, biophysical techniques that include spectroscopy methods (X-ray crystallography, UV-Vis, IR, fluorescence, and others), surface plasmon resonance (SPR), isothermal titration calorimetry (ITC), and nuclear magnetic resonance (NMR) can help to validate the action sites of molecular targets of natural compounds. These tools have been used for decades in drug design in the pharmaceutical industry since they allow a detailed mechanistic characterization of compound binding. This information is quite useful when studying the mechanisms of action of natural plant compounds. Therefore, this chapter provides an overview of biophysical techniques that can validate targets of phytochemicals with anti-virulence properties that can help elucidate their mode of action.

Chapter 3 - Pathogenic bacteria outbreaks cause international economic losses and public health issues during food production and clinical environments. New non-toxic and efficient treatments for disinfecting are highly demanded these days. In this context, terpenes are natural compounds widespread in plants and have attracted attention due to their antibacterial and anti-virulence potential. This chapter discusses the terpenes' anti-virulence mode of action, inhibition of biofilm, disruption of the cell wall and bacterial membrane, intercellular leakage, interruption of nucleic acids transcription, and interference with the quorum sensing signaling process. Overall, promising applications of terpenes compounds in the clinical and food industry as natural antibacterial treatments that effectively reduce pathogenicity and control resistant bacteria are evidenced.

Chapter 4 - Asafoetida is a plant being used since hundreds of years for edible as well as medicinal purposes. The authors undertook to investigate anti-pathogenic potential of essential oil from this plant obtained through microwave-assisted extraction method. This oil was found to be able to inhibit growth and quorum sensing-regulated pigment formation of pathogenic bacteria like *Pseudomonas aeruginosa* and *Chromobacterium violaceum*. This oil (from Kabuli variety) was able to reduce virulence of *P. aeruginosa* towards the model host *Caenorhabditis elegans*. Repeated exposure of *P. aeruginosa* to asafoetida oil was not found to induce any resistance in this bacterium against this oil's anti-infective activity. *Iran, Shiro,* and *Uzbeki* varieties of asafoetida were able to inhibit growth of *Shigella flexneri, Staphylococcus hominis,* and *Vibrio cholerae*. Various phenotypic traits (e.g., biofilm formation, antibiotic susceptibility, and catalase activity) of some of these pathogenic bacteria were also modulated under influence of asafoetida oil. Though the authors' hitherto experiments indicate asafoetida oil to possess anti-pathogenic activity, in most cases, such activity is exhibited at relatively higher concentrations, and hence it seems necessary to isolate the bioactive fractions/compounds from this oil for their individual assessment with respect to potential antibacterial activity.

Chapter 5 - Effect of audible sound waves on bacteria has remained a poorly investigated area. The authors have been investigating microbial response to sonic stimulation for the last several years. This chapter describes a part of the authors' work in this field, wherein different mono-frequency and poly-frequency sounds were investigated for their possible therapeutic effect on infected worm (*Caenorhabditis elegans*) population. Sound corresponding to the frequency of 400 Hz, 700 Hz and 2,000 Hz were found to confer 11-31% survival benefit on worm population challenged with multi-drug resistant gram-negative pathogen *Serratia marcescens*. A combination of sound pattern and the antibiotic chloramphenicol was found to be more effective at saving worms from bacterial attack than either sound or antibiotic alone. Further investigations to elucidate the mechanistic details are warranted.

In: Pathogenic Bacteria
Editor: Keith D. Watts
ISBN: 978-1-68507-422-7

Chapter 1

INHIBITION OF EXTRACELLULAR POLYMERIC SUBSTANCES TO REDUCE BACTERIA PATHOGENICITY

L. A. Ortega-Ramirez[1], M. M. Gutiérrez-Pacheco[1], J. C. López-Romero[2] and J. F. Ayala-Zavala[3,*]

[1]Departamento de Ciencias de la Salud, Universidad Estatal de Sonora. San Luis Río Colorado, Sonora, México

[2]Departamento de Ciencias Químico-Biológicas y Agropecuarias, Universidad de Sonora, Unidad Regional Norte, Caborca, Sonora, México

[3]Coordinación de Tecnología de Alimentos de Origen Vegetal, Centro de Investigación en Alimentación y Desarrollo, A. C. Carretera Gustavo Astiazarán, Colonia la Victoria, Hermosillo, Sonora, México

---

[*] Corresponding Author's E-mail: jayala@ciad.mx.

ABSTRACT

The persistence and resistance of bacteria to disinfection processes are associated with their ability to form biofilms on biotic and abiotic surfaces. Bacteria adhere to natural and engineered surfaces and become mature biofilms encased in self-produced extracellular polymeric substances (EPS). EPS consists of polysaccharides, proteins, metabolites, and nucleic acids. It is well known that cells in biofilms are significantly more resistant to disinfectants and external attacks. It should be noted that EPS may also be involved in cell defense. One of the main functions of EPS is to prevent the penetration of dangerous molecules into bacterial cells, interacting with chemicals making them non-bioavailable, and degrading/hydrolyzing dangerous compounds. This characteristic makes EPS potentially dangerous in the field of food and medicine. Exopolysaccharides are formed in the biofilm of many bacterial species such as *Escherichia* spp., *Salmonella* spp., *Bacillus* spp., *Pseudomonas* spp., *Klebsiella* spp., and *Enterobacter* spp. In the case of *Salmonella*, cellulose is one of the main polymeric substances produced by the cellulose synthase enzymatic complex, made up of glycosyltransferases. Therefore, a mechanism for controlling biofilms can be inhibiting this enzyme, and consequently, reducing pathogenicity. The EPS of *P. aeruginosa* predominantly consists of different polysaccharides and proteins. Mucoid strains of *Pseudomonas* spp. are characterized by the overproduction of alginate, an extracellular polysaccharide, increasing its pathogenicity. Alginate is an important virulence factor in chronic lung infections in patients with the inherited disease cystic fibrosis, and it is the main reason for cystic fibrosis patients' premature death. Compounds obtained from natural sources, enzymatic treatments, chemical synthetic, and nanoparticles have shown effectiveness in inhibiting EPS of pathogenic bacteria. Several essential oils can inhibit the formation of biofilms, and this effect was related to the inhibition of the EPS production mechanism. In general, EPS as mediators of biofilm formation and stability should be considered target structures for corrective actions to eliminate, prevent, or control undesirable biofilms and bacteria-influenced corrosion of materials, both in the food industry and in the field of medicine. Therefore, the objective of this chapter is to describe the virulence problem caused by EPS and antibacterial compounds capable of inhibiting their production, biofilm formation, and bacterial pathogenicity.

Keywords: biofilm, EPS, virulence, antibacterial, infection

INTRODUCTION

The persistence and resistance of bacteria to disinfection processes are associated with their ability to form biofilms on biotic and abiotic surfaces. Bacteria adhere to natural and engineered surfaces and become mature biofilms encased in self-produced extracellular polymeric substances (EPS). The bacteria's persistence is associated with their ability to adhere strongly to surfaces (such as plastics, metals, medical devices, plants, and body tissues), multiply, and form biofilms. Biofilms are communities of bacteria enclosed in a self-produced matrix of EPS such as polysaccharides, lipids, lipoproteins, proteins, and DNA, whose composition can vary among bacteria (Gutiérrez-Pacheco, Bernal-Mercado et al. 2019). It has been reported that the EPS matrix, also called matrixome, promotes the strong adhesion of bacteria to surfaces. Also, it provides a three-dimensional network that creates a microenvironment that is fundamental to biofilm lifestyle and virulence because it protects bacterial cells from environmental stresses such as antimicrobials and immune system cells (Karygianni, Ren et al. 2020). Biofilm formation, composition, and incidence vary in different environments, dependent on bacterial species, temperatures, humidity, and substrates. Mainly, the food industry and healthcare places have been commonly associated with biofilm-associated infections.

Food contamination by pathogenic bacteria is considered an emergent public health issue around the world. Foodborne diseases increase every year, being about 600 million people getting sick and 420,000 dying annually after eating contaminated food (WHO 2020). The most common pathogens found in contaminated foods and food industry surfaces are *L. monocytogenes*, *Salmonella* species, and *E. coli*, depending on the industry type. Different disinfectants are used to reduce the microbial load of surfaces; however, these compounds become inefficient or high concentrations are required to achieve this effect (Joshi, Mahendran et al. 2013). This effect is attributed to bacteria being found in the environment as part of bacterial aggregates called biofilms, which are more resistant to antimicrobial agents than their planktonic counterparts (Byun, Han et al.

2021). For this reason, bacteria persist, and cases of cross-contamination continue to appear and cause foodborne illness.

Both food and food processing surfaces are ideal environments for biofilm formation because they combine a perfect substratum, oxygen concentration, moisture, temperature, and hydrodynamic effects (Pagán and García-Gonzalo 2015). Several authors have studied the influence of organic material in the growth of biofilms. For example, Li, Feng et al. (2017) reported that the addition of chicken meat exudates in a growth medium make *Campylobacter coli* and *Campylobacter jejuni* form more biofilm on glass, polystyrene, and stainless steel surfaces (Brown, Reuter et al. 2014), commonly materials found in the food industry as part of work tables, equipment, and tools. A similar effect was observed with meat juice, facilitating biofilm formation of *Campylobacter* and *Salmonella* under both static and flow conditions. Additionally, to nutrients and organic matter, surface characteristics such as composition and topography became important factors that promote biofilm development (Aryal and Muriana 2019).

Studying the characteristics of the bacterial aggregates is important to ensure the efficiency of the disinfection processes and consequent assurance of food safety. Biofilm formation is a process in which bacteria undergo a nomadic unicellular to sessile multicellular lifestyle change (Romeo 2008). Previous studies allow the construction of a hypothetical developmental model for biofilm formation that has been generalized for different bacterial species (Romeo 2008). The biofilm formation process begins with adherence to a surface. Once adhered, it begins to divide and the daughter cells spread around the attachment site. In a later stage, the bacterium begins to secrete EPS, which constitutes the biofilm matrix. Finally, some bacteria of the biofilm matrix are released to colonize new surfaces (Kishen et al. 2010, Kerekes et al. 2013). In general, EPS as mediators of biofilm formation and stability should be considered target structures for corrective actions to eliminate, prevent, or control undesirable biofilms and bacteria-influenced corrosion of materials, both in the food industry and in the field of medicine.

Therefore, the objective of this chapter is to describe the possible antibacterial compounds capable of inhibiting the production of EPS and consequently the formation of biofilms, which would reduce pathogenicity.

Problems Associated with Biofilm Formation

Bacterial adhesion to biotic and abiotic surfaces is an issue of considerable concern facing different areas such as the food industry and health environments. It has been associated with increased infections and high economic losses (Khelissa, Abdallah et al. 2017). The persistence of bacteria in these places is associated with their adherence ability (in plastics, metals, medical devices, plants, and body tissues), which multiply, and form aggregates called biofilms. Biofilms are communities of bacteria enclosed in a self-produced matrix of extracellular polymeric substances (EPS) such as polysaccharides, lipids, lipoproteins, proteins, and DNA, whose composition can vary among bacteria (Gutiérrez-Pacheco, Bernal-Mercado et al. 2019).

It has been reported that the EPS matrix promotes the strong adhesion of bacteria to different surfaces. Also, it provides a three-dimensional network that creates a microenvironment that is fundamental to biofilm lifestyle and virulence because it protects bacterial cells from environmental stresses such as antimicrobials and immune system cells (Karygianni, Ren et al. 2020). Biofilm formation, composition, and incidence vary in different environments, dependent on bacterial species, temperatures, humidity, and substrates. Mainly, the food industry and healthcare places have been commonly related to biofilm-associated infections.

Biofilms in the Food Industry

Food contamination by pathogenic bacteria is considered an emergent public health issue around the world. Foodborne diseases increase every year, being about 600 million people getting sick and 420,000 dying annually after eating contaminated food (WHO 2020). The Centers for Disease Control and Prevention (CDC 2021) reported on September 1 of the current year six active investigations of food outbreaks linked to *Listeria monocytogenes* (fully-cooked chicken), *Salmonella* (Italian-style meats, packaged salad greens, raw frozen stuffed chicken and frozen cooked shrimp, as well as Shiga toxin-producing *E. coli* (cake mix). These infections caused a total of 38 hospitalizations, 103 illness people, and one death; however, from January to today, ten multistate outbreaks investigations were reported. The most common pathogens found in contaminated foods and food industry surfaces are *L. monocytogenes*, *Salmonella* species, and *E. coli*, which can vary depending on the industry type (Table 1). They may come from any step in the farm-to-fork line from environmental, animal, or human contamination sources (Mritunjay and Kumar 2015). For example, fresh produce is contaminated at pre-harvest by using irrigation water contaminated with human and animal feces, the presence of animals in the field, the contact of produce with the soil, the use of animal wastes for fertilization, and the unhygienic handling of farmers (Rajwar, Srivastava et al. 2016).

Milk and other dairy products are often contaminated as a cause of sick cows, poor animal and milk hygiene practices (Velázquez-Ordoñez, Valladares-Carranza et al. 2019). Whereas meat and meat products are contaminated with pathogens found in the feed, pastures, skin, or during the slaughtering from enteric pathogens in fecal matter or by contact of carcasses with contaminated hands of workers or equipment (Das, Nanda et al. 2019). All these practices provide part of the inocula that contaminate the subsequent industry processing steps.

Table 1. Biofilm-forming bacteria were found in different food industries

Food industry	Bacteria	References
Fresh produce	*Listeria monocytogenes*	(Prado-Silva, Cadavez et al. 2015, CDC 2021)
	Escherichia coli	(Prado-Silva, Cadavez et al. 2015)
	Salmonella spp.	(Prado-Silva, Cadavez et al. 2015)
	Campylobacter spp.	(Weerasooriya, McWhorter et al. 2021)
	Pseudomonas fluorescens	(Wickramasinghe, Hlaing et al. 2020)
	Pseudomonas lundensis	(Wickramasinghe, Hlaing et al. 2020)
Meat	*Pseudomonas aueroginosa*	(Khanashyam, Shanker et al. 2021)
	Listeria monocytogenes	(Panebianco, Giarratana et al. 2021)
Dairy	*Pseudomonas fragi*	(Meng, Zhang et al. 2017)
	Pseudomonas fluorescens	(Meng, Zhang et al. 2017)
	Listeria monocytogenes	(Panebianco, Giarratana et al. 2021)
Fish	*Shewanella putrefasciens*	(Tan, Li et al. 2021)
	Vibrio parahaemolyticus	(Tan, Li et al. 2021)
	Vibrio cholerae	(Ashrafudoulla, Na et al. 2021)

The major contamination source in food industry plants derives from the incoming raw materials, containing spoilage and pathogenic bacteria retrieved from the pre-processing environment as mentioned above, and during different processing steps (Manios, Kapetanakou et al. 2014). These include handling, washing, cutting, storage, packaging, transportation, and others, and are commonly associated with poor hygiene practices (Muhterem-Uyar, Dalmasso et al. 2015). Particularly in fresh produce, soil in the crop or other organic matter inactivates the disinfectants used during the washing and disinfection when the products arrive at the processing plant (Gil, Selma et al. 2015). Furthermore, the water used for making ice or the spraying for cooling is also a risk to consider because it could have poor microbiological quality, enabling pathogenic microorganisms to contaminate foodstuffs (Gil, Selma et al. 2015). This microbial load may be easily spread to abiotic or biotic surfaces or even to the final product facilitating cross-contamination (Manios, Kapetanakou et al. 2014).

Different disinfectants are used in the food industry to reduce the microbial load of surfaces; however, these compounds become inefficient or high concentrations are required to achieve this effect (Joshi,

Mahendran et al. 2013). This effect is attributed to bacteria being found in the environment as part of biofilms, which are more resistant to antimicrobial agents than their planktonic counterparts (Byun, Han et al. 2021). For this reason, bacteria persist, and cases of cross-contamination continue to appear and cause foodborne illness. Both food and food processing surfaces are ideal environments for biofilm formation because they combine an ideal substratum, oxygen concentration, moisture, temperature, and hydrodynamic effects (Pagán and García-Gonzalo 2015).

Biofilm in the food industry often colonizes equipment, surfaces (both food-contact and non-food-contact), and the final food product. Biofilm formation in the food industry is favored by the precipitation and establishment of organic molecules from foods on the adhesion surface (conditioning film) (Shi and Zhu 2009). According to the processed food, one of the main factors that promote biofilm formation is the different organic materials and nutrients. Therefore, the adsorption of organic material onto a surface may be of particular importance in the subsequent development of biofilm architecture. For example, in the fish industry it is most common to find muscle proteins troponin, tropomyosin, and myosin, and lipids, whereas in the dairy industry, surfaces casein, lactalbumin, and some types of lipids. In the fresh produce industry, mainly sugars and vitamins. This composition will affect cell retention and subsequent biofilm architecture, function, and cleanability (Hingston, Stea et al. 2013).

Several authors have studied the influence of organic material in the growth of biofilms. For example, Li, Feng et al. (2017) reported that the addition of chicken meat exudated in a growth medium made *Campylobacter coli* and *Campylobacter jejuni* form more biofilm on glass, polystyrene, and stainless steel surfaces (Brown, Reuter et al. 2014), commonly materials found in the food industry as part of work tables, equipment, and tools. A similar effect was observed with meat juice, facilitating biofilm formation of *Campylobacter* and *Salmonella* under both static and flow conditions. Additionally to nutrients and organic matter, surface characteristics such as composition (biotic or abiotic), material (stainless steel, plastic, rubber, woods), and topography, became important factors that promote biofilm development (Aryal and Muriana 2019).

In general, biofilm formation improves bacterial survival to the environmental conditions in the food industry, such as refrigeration, acidity, salinity, and disinfectants. The surfaces where biofilms have been identified include food products, tools and equipment (stainless steel tables, knives, pipes, conveyors belts, floors, drains, and storage boxes), water distribution system that could further contribute to food spoilage and cross-contamination, facilitating the spread of foodborne pathogens (Zhao, Zhao et al. 2017, Faille, Cunault et al. 2018).

Fresh Produce

Food-borne infections caused by fresh fruit and vegetables are among the most severe issues worldwide (Amrutha, Sundar et al. 2017). Several bacteria have been associated with fresh produce contamination and spoilage; the most common pathogens include *Salmonella*, *E. coli*, *Listeria, Shigella,* and the spoilage *Pseudomonas* and *Erwinia carotovora* (CDC 2021). These biofilm-forming bacteria have been detected in different processing steps over biotic and abiotic surfaces in the fresh produce industry. Their biofilm formation capacity is influenced by several factors, among which nutrients availability is one of the most important. Han, Klu et al. (2017) compared the biofilm formation of Salmonella species and enterohemorrhagic *E. coli* on biotic (alfalfa and bean sprouts) and abiotic (polystyrene) surfaces. It was observed that *Salmonella* Enteritidis attached better to bean sprouts (5.58 log CFU/g) than alfalfa (5.28 log CFU/g), whereas *E. coli* formed more biofilm on polystyrene surfaces.

L. monocytogenes, naturally occurring in agricultural environments such as soil, manure, and water, have been implicated in multistate fresh produce outbreaks. Their biofilm formation in fresh produce industry surfaces has been evaluated in the presence and absence of lettuce leaf extracts to assess the influence of nutrients. It was found that at 10°C, the biofilm counts increased from 3 to 6.4–7.2 log CFU/cm^2, whereas at 4°C, the biofilm formation increased to 4.3–4.8 log CFU/cm^2 after 240 h incubation (Kyere, Foong et al. 2020). *E. coli* is one of the main pathogenic bacteria associated with fresh produce outbreaks; particularly, Shiga toxigenic *E. coli* (STEC) was commonly associated with these infections. These are characterized by cause diarrhea, hemorrhagic colitis, and hemolytic uremic syndrome, causing acute renal failure in children and immune-compromised people (Griffin and Karmali 2016). Most human STEC infections are associated with serotype O157:H7, but awareness of the public health importance of non-O157 serotypes is growing. In this sense, several studies have been conducted to evaluate the biofilm formation of this bacteria in the food industry environment.

Adator, Cheng et al. (2018) evaluated the biofilm formation of 14 serotypes of the STEC on stainless steel and polystyrene surfaces, as well as their ability to survive after 30 days. Results demonstrated that in only 2 minutes of contact between lettuce and biofilm contaminated surfaces, high counts of E. coli strains were transferred. For example, the 6-old biofilm of O157-508 shared 6.5 log CFU/g, whereas at two days, the counts in lettuce of serotypes O145-099, O157-R508, and O111-CFS, were 6.35, 6.48, and 5.94 log CFU/g, respectively. Concerning stainless steel, 1.42 log CFU/g of *E. coli* O157:H7-R508 were transferred to lettuce, followed by the strains CFS and 053 of E. coli O111 at 0.60 log CFU/g. It is important to mention that all STEC strains survived for over a month in polystyrene and stainless-steel surfaces, highlighting the serious problem of the persistence of biofilms in the fresh produce industry and the ease and speed of cross-contamination.

Seafood

The common seafood bacterial pathogens that form biofilms in the fish industry are *Aeromonas hydrophila, Vibrio* spp., *Salmonella* spp., and *L. monocytogenes*, which causes significant health and economic issues (Mizan, Jahid et al. 2015). Contamination of seafood products is mainly associated with pathogens in water and the gastrointestinal tract of animals, which causes cross-contamination during harvesting, handling, processing, transportation, and storage. The constant cases of food outbreaks are linked to the consumption of contaminated raw or undercooked seafood.

Vibrio is one of the main bacterial genera associated with seafood infections, and it includes *Vibrio cholerae*, *Vibrio harveyi*, *Vibrio parahaemolyticus*, *Vibrio vulnificus*, and *Vibrio alginolyticus*, which can form biofilms on seafood as well as on surfaces that come into contact with food (Roy, Mizan et al. 2021). *V. parahaemolyticus* is a foodborne pathogen that causes acute gastroenteritis and is generally isolated from different kinds of seafood such as shrimp, fish, oysters, and crabs. It was reported that *V. parahaemolyticus* isolates formed biofilms on seafood industry surfaces such as stainless steel (5.39 log CFU/cm^2), high-density polyethylene (5.9-6.0 log CFU/cm^2), glass (5.9 log CFU/cm^2), and on the operculum of *Micropogonias furnieri* (Whitemouth Croaker) (5.7-5.9 log CFU/cm^2). The presence of biofilms on abiotic surfaces limited the disinfection potential of sodium hypochlorite, being 20 parts per million (20 ppm) insufficient to eliminate *V. parahaemolyticus* cells (Rosa, Conceição et al. 2018). On the other hand, Wang et al. (2018) evaluated the biofilm formation of the spoilage bacteria *Pseudomonas fluorescens*, commonly found in dairy, poultry, and fresh produce industries. Results showed that after just 10 min of exposure, 4.5 log CFU/cm^2 cells were attached to stainless steel surfaces and continued to increase, reaching 5.4 log CFU/cm^2 after five h. Additionally, *P. fluorescens* showed a high cell surface hydrophobicity indicating high adhesion potential.

E. coli is another pathogenic bacterium commonly isolated from the gastrointestinal tract of fish and shellfish (Balière, Rincé et al. 2015). Thus, incorrect evisceration processes and manipulation must lead to cross-contamination of unaffected surfaces and fish products. The contamination sources, biofilm-forming ability, and resistance to disinfection of *E. coli* O157:H7 and non-O157 isolated from tilapia processing plants were evaluated in a study. Results demonstrated that the major counts of *E. coli* were found on frozen tilapia fillets, followed by nitrile and cotton gloves (from slaughtering and filleting, respectively), knives, bleeding tanks, among others. All strains form more biofilm on polystyrene than stainless steel, reaching counts of 5.25 log CFU/cm^3 at the first five h and around

8.09 log CFU/cm^3 at 48 h. The high biofilm formation on these surfaces caused resistance to peracetic acid and sodium hypochlorite, requiring higher doses than those recommended by manufacturers (Vázquez-Sánchez, Antunes Galvão et al. 2018).

Dairy Products

Several pathogens in the dairy industry can form biofilms in static or flow conditions. For this reason, it can be created at different temperatures and with diverse colonizing species on milk and cheese tanks, pipelines, pasteurizers, and packing tools, act as surface substrates for biofilm formation (Galie, García-Gutiérrez et al. 2018). Biofilm forming *Pseudomonas* species such as *P. fluorescens, Pseudomonas lundensis*, and *Pseudomonas fragi* are among the most important bacteria causing spoilage at low storage temperatures in milk products due to the secretion of lipolytic and proteolytic enzymes secreted into raw milk. These enzymes can survive the pasteurization process and reduce the sensory quality and shelf-life of the processed liquid milk products (Meng, Zhang et al. 2017). Meng, Zhang et al. (2017) analyzed raw milk samples collected from 87 bulk tanks of 87 farms in Shaanxi Province in China. The authors obtained 143 isolates from the collected samples and confirmed the presence of 14 biofilm-forming *Pseudomonas* species. Furthermore, proteolytic activity was observed in 40 isolates.

Milk and other dairy products have been contaminated by the biofilm and toxin-producing bacteria *Bacillus cereus*. This bacterium is a common foodborne pathogen in the dairy industry; it can form different types of biofilms both under static and flow conditions, and it has been reported that biofilm matrix serves as a favorable niche for emetic toxins accumulation, which is the reason for their associated gastrointestinal infections (Huang, Flint et al. 2020). Several studies reported the presence of this bacteria in products of different regions of China, such as pasteurized milk (Gao, Ding et al. 2018) and infant formula (Yang, Yu et al. 2017). One of the main problems associated with *B. cereus* biofilms is their resistance, remaining active even after pasteurization or ultra-high temperature (UHT) sterilization (Gopal, Hill et al. 2015).

Meat

The reported foodborne pathogens in the meat processing industry include *L. monocytogenes, E. coli, C. jejuni, Salmonella spp, Pseudomonas aeruginosa,* and *Staphylococcus aureus*. They have been associated with contamination and spoilage of fresh, cured, and ready-to-eat meat products due to their ability to adhere to food contact surfaces and form biofilms. Researchers evaluated the prevalence of *L. monocytogenes* on food industries surfaces for ready-to-eat processing meat. They found that the prevalence of *L. monocytogenes* is 25% higher, even after implementing good hygiene and manufacturing practices. The presence of this bacterium was associated with the handling of the product after processing and hygiene procedures due to the *L. monocytogenes* biofilms resistance (Henriques, Telo da Gama et al. 2014).

On the other hand, Bhardwaj, Taneja et al. (2021) were able to identify an antimicrobial-resistant biofilm-forming *E. coli* pathotype isolated from dairy and meat products. EMC17, an *E. coli* isolate, was established as a powerful biofilm former that reached maximum biofilm formation in 96 h on glass and stainless-steel surfaces. The presence and expression of genes associated with virulence, such as adhesins, invasins, and polysaccharides, were detected. EMC17 turned out to be a multi-resistant strain, possessing extended-spectrum β-lactamases and a biofilm phenotype. Early production of *quorum sensing* molecules (N-acyl homoserine lactones) and EPS production facilitated the early initiation of biofilm formation by EMC17.

Furthermore, the genes that make up the EMC17 biofilm were significantly upregulated from 3 to 27 folds in the biofilm state. Bacteria such as *C. jejuni, E. coli,* and *S. aureus* can interact and improve their survival during the processing and storage of retail meat. These bacteria are highly prevalent in meat foods such as chicken meat, chicken liver, beef liver, pork meat, and turkey meat. A study by Karki, Ballard et al. (2021) found that viable *S. aureus* cells and filter sterilized cell-free media obtained from *S. aureus* prolonged the prevalence of *C. jejuni* at low temperature and during aerobic conditions; this was due to the presence of *S. aureus,* which induced the formation of biofilms of *C. jejuni*.

Biofilms in the Healthcare Environment

Biofilms are a critical issue in the healthcare environment as they can adhere to host tissue and an implantable medical device and can cause many chronic infections. One of the problems associated with biofilms is antibiotic resistance since bacteria within the polymeric matrix are generally unaffected by antibiotics and the human immune system. This response is attributed to EPS limiting chemicals from the surrounding environment into the matrix (Kumar, Chandra et al. 2019). Notably, bacterial biofilm formation is increasingly recognized as a passive virulence factor facilitating many infectious disease processes. Facultative human pathogens can form biofilms outside the host as a model of persistence. As mentioned above, biofilm formation facilitates environmental survival and allows a high infectious dose to be maintained even during long periods between epidemics. That is, biofilm communities represent a reservoir for future infections. After infection, bacteria in biofilms are better protected against host defense mechanisms than their planktonic counterparts. Therefore, biofilms can be considered a feasible way in which bacterial pathogens initiate infection of a human host (Hancock, Alford et al. 2021).

The oral microbiome is one of the most diverse among bacterial habitats in various human body parts and all populations. Disruption of this balance and genetic factors cause oral diseases, including caries, caused by a mixture of microorganisms and food debris and the most common cause of dental pain and tooth loss. *S. mutans* have been linked to caries due to their virulence, determined mainly by their interaction with other less virulent bacteria in the oral microbiome (AlEraky, Madi et al. 2021). Dental caries is considered one of the major chronic non-communicable diseases by the World Health Organization. For tooth decay to appear, the presence of bacteria is necessary. *S. mutans* is one of the main cariogenic bacteria. These bacteria usually exist in the form of biofilms to adapt to the changing oral environment. Typical virulence phenotypes of *S. mutans* are acid production, acid resistance, adhesion, intracellular and extracellular polysaccharide synthesis as glucans for forming its biofilm, allowing it to retain nutrients and protect itself from antibacterial agents (Yan, Wu et al. 2020).

Another disease caused by the formation of bacterial colonies is periodontitis, an infection of the gums that damages the soft tissues and bones that support the teeth. *Fusobacterium nucleatum* and *Pseudomonas aeruginosa* are the cause of this periodontal infection. These microorganisms colonize the mucosa of the oral cavity, altering the flow of calcium, invade the mucosa cells, and release toxins, and 15 days later, plaque formation appears, which is also a community of biofilms and the final cause of periodontitis (Gupta, Sarkar et al. 2016). The absorption of hydrophobic macromolecules produces oral biofilm to the tooth surface, which will form the acquired film. The main bacteria that initially adhere are *S. mutans* and *Lactobacillus* using the adhesion proteins Pac and glucosyltransferase. Then, sucrose degradation gives rise to the extracellular matrix interacting by special electrostatic forces other types of bacteria such as *Prevotella sp.* and *Actinomycetes*, causing adhesion of these to the newly formed extracellular matrix (Magennis, Francini et al. 2017).

Gastrointestinal infections have been mainly associated with consuming food or water contaminated with feces and exposure to pathogens in the hospital environment (Negrut, Khan et al. 2020). Annually, the CDC reports foodborne illnesses, causing around 4200 hospitalizations from 2018 up to date. The most common symptoms of gastrointestinal diseases caused by pathogenic microorganisms include diarrhea, vomiting, fever, headache, and even death, depending on the microorganism causing the infection (CDC 2021). Gastrointestinal infections continue to cause illness and death and contribute to economic loss in most parts of the world (Ternhag, Törner et al. 2008). It has been estimated that they are the second most common infectious diseases after respiratory tract infections and a major cause of morbidity and mortality among infants and children (WHO 2017).

Gastrointestinal disorders can be caused by many microorganisms such as viruses, protozoa, fungi, and bacteria, the most common causal agents. The human gastrointestinal tract encompasses many nutritional and physicochemical environments, ideal for biofilm formation and survival (von Rosenvinge, O'May et al. 2013). *V. cholerae* causes a waterborne diarrheal disease called cholera; its rapid colonization of the gastrointestinal tract is due to its ability to form biofilms in aquatic environments on chitinous surfaces and the human host (Schulze, Mitterer et al. 2021). It is important to mention that biofilms cause infections, but the cells shed from the biofilm become much more infectious. The impact of biofilms on *V. cholerae* transmission was evaluated, highlighting that by filtering the water with a cloth, cholera cases were reduced by 50%. This observation confirms that water in contact with bacterial aggregates or mature biofilms is more likely to be infected with *V. cholerae*, highlighting the ecological and epidemiological role of the biofilm (Colwell, Huq et al. 2003).

Infectious diarrhea caused by outbreaks of *Salmonella* species is the most likely cause in children from industrialized countries (Uddin, Rahman et al. 2021). Salmonellosis has increased in recent years in the United States, and it is an important cause of diarrheal disease; according to CDC (2021) data, there have been more than 10,000 cases in the last five years by different species of *Salmonella*. In September 2021, a S*almonella* outbreak was linked to frozen cooked shrimp, causing nine people to become ill. Most people infected with *Salmonella* experience diarrhea, fever, and stomach cramps. Symptoms generally begin 6 hours to 6 days after ingestion of the bacteria (CDC 2021). The ability of *Salmonella* to form biofilms contributes to its resistance and persistence in both host and non-host environments and is especially important in food processing settings (Steenackers, Hermans et al. 2012). A study found that a *Salmonella* rdar morphotype could be an adaptation strategy of *Salmonella* virulence (this morphotype only is expressed under specific environmental conditions). The main reason for its persistent nature is producing an abundant extracellular matrix formed by proteinaceous compounds consisting of adhesive curli fimbriae or thin aggregating fimbriae (Steenackers, Hermans et al. 2012).

Globally, respiratory infections are the main cause of morbidity and mortality from infectious diseases worldwide. More than one billion people are affected by acute and chronic respiratory diseases. These infections mainly affect infants, children, and older people (Murdoch and Howie 2018). They are typically caused by viruses, bacteria, or mixed viral–bacterial infections, can be contagious, and spread rapidly through respiratory droplets. The main bacteria related to respiratory infections are *Haemophilus influenzae*, *Streptococcus pyogenes*, *Streptococcus pneumoniae*, and *Moraxella catarrhalis* (Zhang, Wang et al. 2020). The large surface area of the mucosa makes the respiratory tract a preferred niche for biofilm growth, which can lead to chronic inflammation of the mucosal tissue and reduced lung function (Huffnagle, Dickson et al. 2017).

H. influenzae is the most common cause of bacterial infection in the lungs of patients with chronic obstructive pulmonary disease. Due to its ability to adhere to host epithelial cells, initial colonization of the lower respiratory tract can progress to persistent infection and biofilm formation (Short, Carson et al. 2021). It is characterized by changes in bacterial behavior, reduced cell metabolism, and the production of an obstructive extracellular matrix. Biofilms of *H. influenzae* use several mechanisms to persist in the host. The production of a robust extracellular matrix aids resistance to neutrophils and proteases. A thick extracellular matrix prevents phagocytosis by blocking macrophage access to biofilm cells, pairing it with reduced cell metabolism and protein synthesis, the activity of antibiotics in biofilm cells is minimal. Another resistance mechanism is eDNA incorporated into the biofilm matrix that binds to β-defensin, reducing its antimicrobial properties, and finally, β-lactam antibiotics induced biofilm formation in *H. influenzae* in response to stress (Short, Carson et al. 2021).

S. pneumoniae is a leading cause of pneumonia mortality globally. Pneumococcal disease is often associated with prolonged colonization of hosts, and this process is facilitated by biofilm formation that is largely resistant to conventional antibiotics (Boswell and Cockeran 2021). *Pneumococcal* biofilms are essential for colonization and persistence in different environmental conditions that affect their formation (Hameed, Motib et al. 2021). The bacterial cells in these biofilms are held together by an extracellular matrix composed of DNA, proteins, and possibly polysaccharides. Although neither the precise nature of these proteins nor the composition of the putative polysaccharides is precise, it is known that choline-binding proteins are required for successful biofilm formation. Additionally, many genes appear to be involved, although the role of each seems to vary when biofilms are produced in different environments (Domenech, García et al. 2012).

M. catarrhalis is a Gram-negative mucosal pathogen of the human respiratory tract. Although little information is available regarding the initial steps of *M. catarrhalis* pathogenesis, this organism must be able to colonize the human mucosal surface to initiate an infection. Type IV pili (TFP), filamentous surface appendages primarily comprised of a single protein subunit termed pilin, play a crucial role in initiating disease by a wide range of bacteria. *M. catarrhalis* cells form a mature biofilm in continuous-flow chambers, and that biofilm formation is enhanced by TFP expression. There have been numerous biological functions associated with TFP expression by various bacterial species. These may include adherence to eukaryotic cells, biofilm formation and stability, competence for natural transformation, and flagellum-independent cell movement (Luke, Jurcisek et al. 2007).

In August 2020, the CDC reported a total of 40 people infected with the outbreak strain of *E. coli* O157:H7, 20 people were hospitalized, and 4 developed hemolytic uremic syndrome, a type of kidney failure (CDC 2020). In *E. coli* strains that cause the hemolytic uremic syndrome, its high virulence is due to the strong adherence typical of enteroaggregative strains, and the formation of biofilms combined with Shiga toxin production. These strains contain diguanylate cyclase (DgcX), which produces the second messenger c-di-GMP that promotes biofilm (Richter, Povolotsky et al. 2014). In addition, it constantly generates derivatives with a production of CsgD and curli. Since curli fibers are strongly pro-inflammatory and cellulose counteracts this effect, the high output of c-di-GMP and curli improve adherence, contribute to inflammation, thus facilitating the entry of Shiga toxins in the bloodstream and the kidneys (Richter, Povolotsky et al. 2014).

Renal infection is a urinary tract infection that usually begins in the urethra or bladder and works up to one or both kidneys. A urinary tract infection is defined as pathogenic microorganisms in any part of the urinary system and represents one of the most common infections worldwide (Bernal-Mercado, Gutierrez-Pacheco et al. 2020). The main microorganisms of urinary infections include *Klebsiella*, *Enterobacter*, *Pseudomonas*, *Acinetobacter*, and uropathogenic *E. coli*, the latter being the main etiological agent. It is well known that the capacity of these microorganisms to adhere and form biofilms on medical devices are the leading causes of urinary infections associated with catheters (Nicolle 2014).

The leading causes of infections of the uterine tract (> 90%) are uropathogenic *E. coli* (UPEC) and *Klebsiella pneumoniae* (UPKP). These infections represent between 10% and 30% of infections in community settings and between 25% and 60% of nosocomial infections, mainly due to catheter-associated infections (Behzadi, Urbán et al. 2020). Genitourinary tract infections are correlated with the ability to form biofilms; for example, UPEC can aggravate illnesses due to its resistance to antibiotics, turning acute conditions into chronic or recurrent infections. Biofilm formation plays a critical role during the pathogenesis of urinary tract infections, particularly in catheter-associated diseases. Since the production of the extracellular matrix helps in the union and protects against the forces of the urinary tract, promoting the persistence and bacterial chronicity (Soto González 2014). Relapse by UPEC has been related to the ability of pathogenic strains to form a biofilm. In a study carried out by Soto, Smithson et al. (2006) in 43 patients with cystitis or pyelonephritis problems, urine cultures were collected for six months, finding 80 strains of *E. coli*, 27 strains caused relapses and 53 reinfections. However, 20 and 22 strains were biofilm producers *in vitro*, demonstrating a relationship between persistence, relapse, and biofilm formation.

Composition of Extracellular Polymeric Substances of Pathogenic Bacteria

The development of bacterial biofilms has been found on practically all surfaces, both in natural and industrial environments. A biofilm is a complex structure made of bacterial aggregates within an EPS matrix. The composition of biofilm is generally microbial cells (2−5%), DNA and RNA (< 1−2%), polysaccharides (1−2%), proteins < 1−2% (including enzymes), and water (up to 97%) as the main component (Brindhadevi, Lewis Oscar et al. 2020). Biofilm formation is when bacteria change from a nomadic unicellular to a sessile multicellular lifestyle, where subsequent growth results in structured communities and cell differentiation (Romeo 2008).

Previous studies allow constructing a hypothetical developmental model for biofilm formation that has been generalized for different bacterial species. The biofilm formation process begins when a high cell load and bacterial adherence appear on the surface (1). Once the bacterium has attached to the surface, it begins to divide, adhering firmly with the secreted EPS (2). The daughter cells spread around the attachment site, forming microcolonies similar to the colony formation process on agar plates (3). Later, the bacterium continues secreting EPS (4) that constitute the biofilm matrix and form channel-like structures until maximum growth (maturation). Finally, some bacteria are released from the biofilm matrix (5) to colonize new surfaces, closing the biofilm formation process (Gutiérrez-Pacheco, Bernal-Mercado et al. 2019). Each of these steps is influenced by several physicochemical and genetic factors (Figure 1). Biofilm formation is a complex and dynamic process in which the EPS of the biofilm matrix plays a critical role in the biofilm's structure and functions (Karygianni, Ren et al. 2020).

The biofilm matrix offers a constantly hydrated viscous layer that protects bacteria from desiccation or host defenses, preventing recognition of bacteria by the immune system (Limoli, Jones et al. 2015, Flemming, Wingender et al. 2016). Furthermore, it could inhibit the washout of enzymes, nutrients, or even signaling molecules that could accumulate locally and create more favorable microenvironments for bacteria within the biofilm (Izadi, Izadi et al. 2021).

EPS synthesis and their spatial organization change between species, but in general, a diverse group of biomolecules can be grouped into two major categories: (i) molecules associated within the cell surface and (ii) those secreted out of the cell system. Examples include cell appendages such as type IV pili, flagella, and functional amyloids that modulate bacterial adhesion, mechanical stability, and autoimmune responses. On the other hand, secreted bacterial proteins, exopolysaccharides, eDNA, and eRNA released extracellularly contribute to matrix scaffolding and function. In general, the different EPS components play an important role in the biofilm, as observed in Figure 2 (Flemming and Wingender 2010). The matrix is a barrier against external agents that can affect the viability of embedded cells, such as some radicals and antimicrobial agents (Gupta, Sarkar et al. 2016). This effect is why biofilms are difficult to remove from the food industry and healthcare environment, causing numerous cases of microbial infections every year (Kyere, Foong et al. 2020). The EPS matrix provides an effective barrier that restricts the penetration of chemically reactive biocides, antibiotics, and antimicrobial agents, making bacteria living in the biofilm matrix up to 1,000 times more resistant to antibacterial compounds than planktonic bacteria (Belfield, Bayston et al. 2015, Flemming, Wingender et al. 2016). Several studies have highly evidenced the resistance exerted by the biofilm EPS matrix to antimicrobial agents (Vázquez-Sánchez, Antunes Galvão et al. 2018).

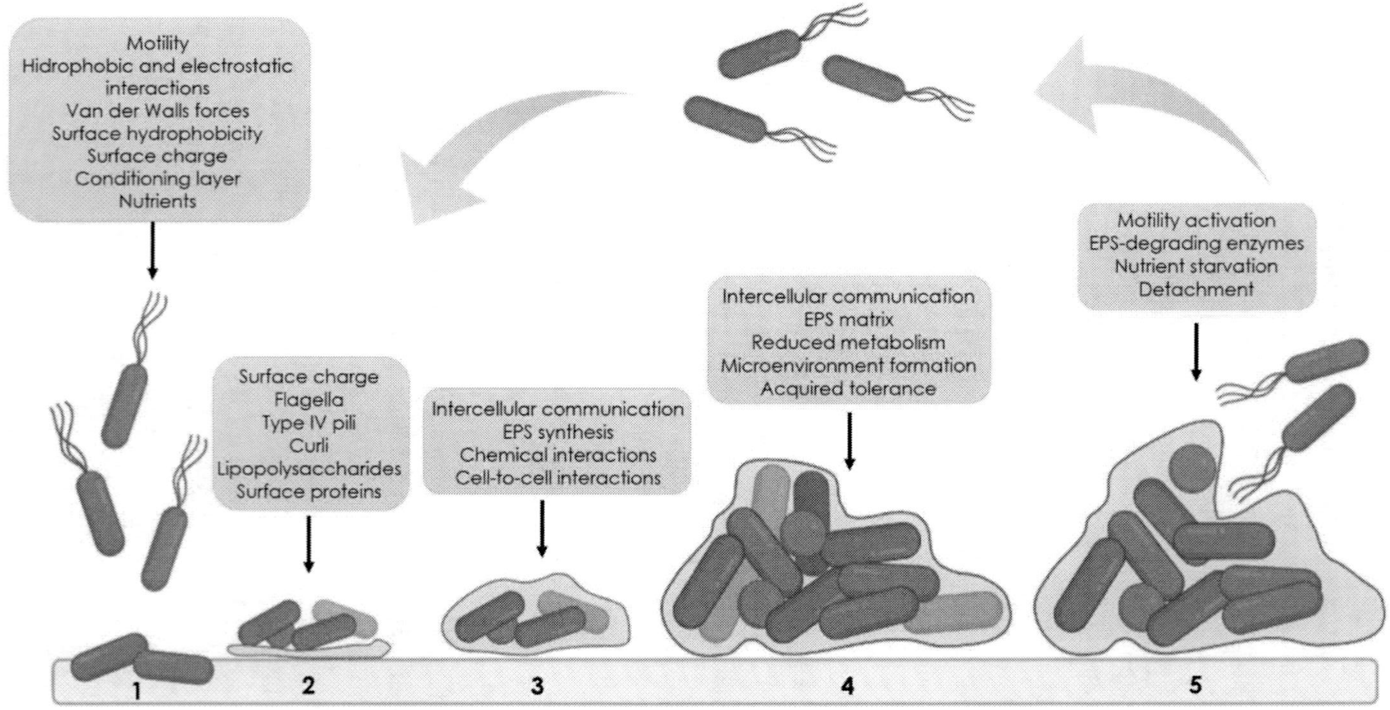

Figure 1. Factors influencing the different steps of the biofilm formation process.

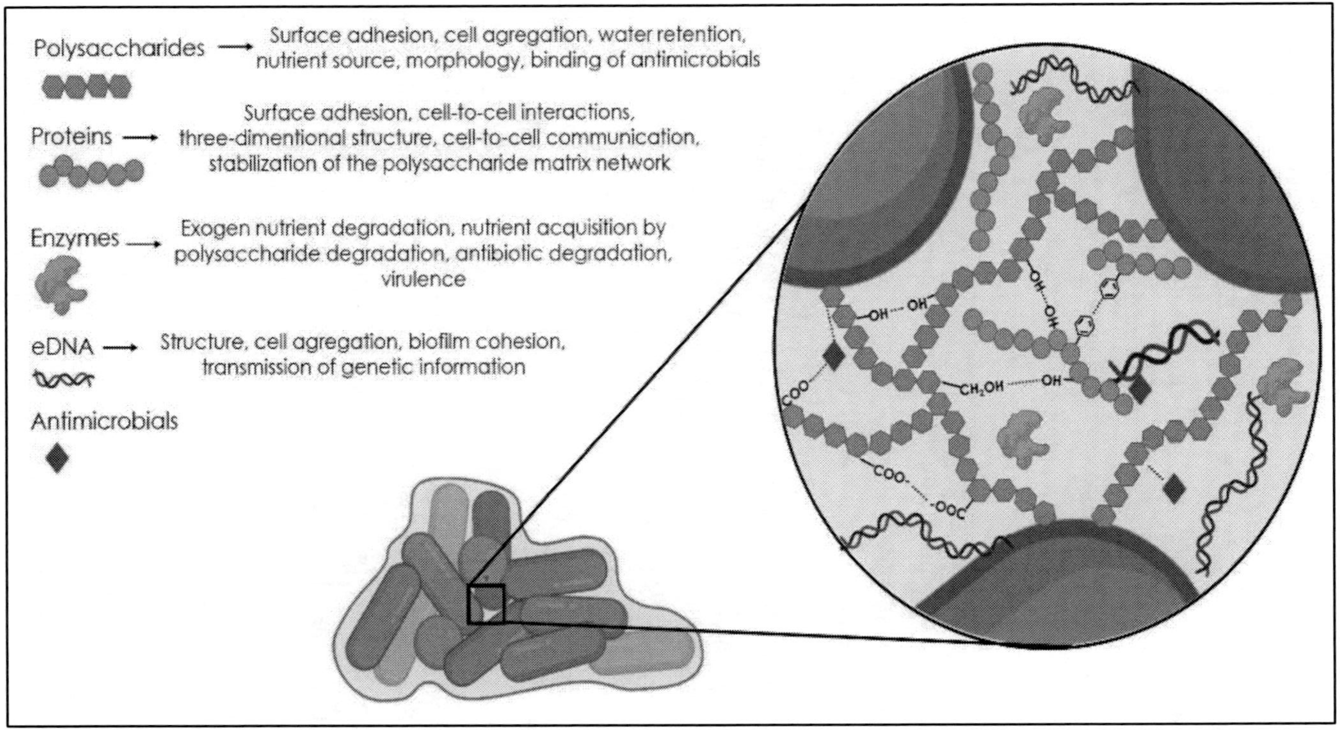

Figure 2. Functions of EPS in the biofilm formation process.

The factors that promote biofilm-specific antimicrobial resistance include the failure to penetrate biofilm, the altered metabolism of biofilm cells, the presence and accumulation of antibiotic-degrading enzymes such as β-lactamases, among others (Olsen 2015). The resistance to antimicrobials in biofilms exerted by the action of antibiotic-degrading enzymes in the matrix has been documented. These were observed in several bacteria such as *K. pneumoniae*, *P. aeruginosa*, *S. aureus*, and *E. coli* (Marques, Motta et al. 2017). *K. pneumoniae* showed resistance to ampicillin due to their β-lactamase activity, confirmed by the deletion of the β-lactamase gene, increasing the bacteria's susceptibility. On the other hand, a study demonstrates that the antibiotic resistance of *S. aureus* biofilms was related to the presence of β-lactamase, as observed by the high bactericidal concentration of cefoxitin (64 μg/mL) and the presence of *icaA* and *icaD* genes that codifies the high-molecular-weight polysaccharide intercellular adhesion (Marques, Motta et al. 2017).

The amount and type of EPS in the biofilm influence important aspects of its architecture. EPS became essential in the adhesion process of bacteria to surfaces and structure development and resistance. EPS reduces the number of antimicrobial molecules that can be transferred to the inner layer of biofilms. The complex structure of the biofilm matrix limits the access of antimicrobials to biofilm cells by steric hindrance or by the direct interaction of EPS components with antimicrobials (Dincer, Özdenefe et al. 2020). For example, Pel and PsI exopolysaccharides in *P. aeruginosa* biofilms can bind antibiotics, providing tolerance to these molecules. A study evaluated the effect of PsI in antibiotic tolerance of *P. aeruginosa* and the non-producing *P. aeruginosa* (mutant Δ*psl), S. aureus, and E. coli.*

Results showed that the presence of PsI makes P. aeruginosa biofilms more resistant to colistin, tobramycin, and ciprofloxacin, showing the Δpsl mutant the lower MIC value (colistin, 3 μg/mL) compared to the PsI producing P. aeruginosa (colistin, 12 μg/mL). Furthermore, it was observed that in the mixed biofilm formation, the presence of PsI increased the antibiotic tolerance of the non-producing PsI cells (*P. aeruginosa Δpsl*, *S. aureus*, and *E. coli*) acquiring a collective resistance as a community (Billings, Ramirez Millan et al. 2013). All these aspects of the putative functions of the matrix contribute to the development of phenotypic resistance of biofilms of pathogenic strains (Karygianni, Ren et al. 2020), and these are dependent on their composition. The chemical structure of EPS is highly diversified, depends on bacteria, genetics, and environmental conditions they grew on, for example, nutrients type (Nouha, Kumar et al. 2018). The microbial species is one of the main factors because they are based on their genetics and metabolic pathways. A single bacterial species can produce several different biofilm matrix components (Flemming and Wingender 2010). The composition of polysaccharides and other components such as proteins often varies between bacteria and even between a single species. However, multiple species of bacteria produce mainly polysaccharides. The most famous exopolysaccharides in biofilms are alginate, cellulose, and poly-N-acetylglucosamine (Table 2).

Pseudomonas aeruginosa is an opportunistic pathogen that causes chronic biofilm infections and can synthesize three exopolysaccharides involved in biofilm formation: alginate, Psl, and Pel. Psl is a neutral polysaccharide consisting of a pentasaccharide repeat-containing glucose, mannose, and rhamnose, and alginate is a negatively charged polymer of guluronic and mannuronic acid. Although previous reports suggest that glucose may be the main component of Pel, its structure is unknown (Jennings, Storek et al. 2015). *E. coli* O157:H7 produces cellulose which is the main exopolysaccharide component of its biofilms. Cellulose is a homopolysaccharide composed of D-glucopyranose units linked by β-1,4 glucosidic bonds and is one of the most abundant polymers on the planet (Zogaj, Nimtz et al. 2001). Cellulose, in addition to having an essential role in the adhesive properties of the matrix (bacteria-surface and bacteria-bacteria interaction) (Beloin, Roux et al. 2008) and as a structural component in the formation of biofilms, also plays an essential role in the resistance of this microorganism to disinfectants. It has been reported that the interaction between cellulose and curli (the main protein component of *E. coli* biofilms) contributes to the three-dimensional growth of the biofilm, as well as to tolerance to desiccation, which could be reflected in the persistence of this bacteria on some surfaces (Flemming, Wingender et al. 2016).

The polysaccharide intercellular adhesin or the poly-N-acetyl glucosamine (PNAG) from the *E. coli* biofilm must adhere to a given surface, increasing virulence (Cerca, Maira-Litrán et al. 2007). PNAG is a positively charged linear homoglycan composed of β-1,6-N acetylglucosamine residues with approximately 20% deacetylated residues (Liang 2015). PNAG forms a protective matrix around bacterial cells involved in cell-cell interactions (Cerca, Maira-Litrán et al. 2007). PNAG can also interact with extracellular DNA by reinforcing the structure of the biofilm matrix (Liang 2015). On the other hand, it was reported that mannose represents around 10% of *E. coli* biofilms (Bales, Renke et al. 2013). In this sense, recent studies have shown that a biofilm can produce different exopolysaccharides depending on the environmental conditions in which it is found (Wang, Ye et al. 2013).

Several studies have reported that EPS plays an important role in pathogenesis; some of these components are virulence factors. A pathogen expresses a virulence factor to influence the vital functions of the host, allowing the pathogen growth and colonization (Gutiérrez-Pacheco, Bernal-Mercado et al. 2019). EPS facilitates the adhesion and colonization surfaces; for this reason, the genetic modification or other mechanisms that alter EPS structure could affect the capacity of bacteria to adhere and colonize, with the concomitant inhibition of biofilm formation. In many bacteria, mutants that cannot synthesize exopolysaccharides are severely compromised or cannot form mature biofilms. In *L. monocytogenes,* a proteomic analysis of the biofilm matrix evidenced the presence of virulence factors called internalins (InlA and InlB). These proteins interact with host-receptors to promote bacterial internalization into epithelial cells during infection and act as cellular adhesins favoring the adhesion process during biofilm formation (Chen, Ross et al. 2011).

Table 2. EPS composition of pathogenic bacteria biofilms

Bacteria	**Extracellular Polymeric Substances**	**Molecule type**	**Function**	**Reference**
Escherichia coli	Cellulose	Polysaccha-ride	Surface adhesion and intercellular aggregation	(Macarisin, Patel et al. 2012)
	Colanic acid	Polysaccha-ride	Biofilm maturation	(Zhang and Poh 2018)
	Curli	Protein	Three-dimen-sional structure and protection	(Ryu and Beuchat 2005)
Pseudomonas aeruginosa	Alginate	Polysaccha-ride	Protection	(Mulcahy, Isabella et al. 2014)
	Rhamnolipid	Lipid	Biofilm structure	(Yu, He et al. 2014)
	pel and psl	Polysaccha-ride	Surface adhesion and structure stability of mature biofilms	(Jennings, Storek et al. 2015) (Billings, Ramirez Millan et al. 2013)
	Lectins (LecA/LecB)	Protein	Surface adhesion	(Passos da Silva, Matwichuk et al. 2019)
Salmonella Typhimurium	Cellulose	Polysaccha-ride	Biofilm structure	(Maruzani, Sutton et al. 2019)
	Curli	Protein	Biofilm structure	(Maruzani, Sutton et al. 2019)
Acinetobacter baumannii	Poly-β (1-6)-*N*-acetylglucosamine (PNAG)	Polysaccha-ride	Intercellular adhesion	(Choi, Slamti et al. 2009)
Staphylococcus aureus	Adhesin (PIA)	Polysaccha-ride	Intercellular adhesion	(Lin, Shu et al. 2015)
	Poly-β (1-6)-*N*-acetylglucosamine (PNAG)	Polysaccha-ride	Intercellular adhesion	(Lin, Shu et al. 2015)
Vibrio cholerae	*Vibrio* polysaccharide (VPS)	Polysaccha-ride	Three-dimensional structure	(Schulze, Mitterer et al. 2021)
	Extracellular DNA	Nucleic acid	Structure	(Seper, Fengler et al. 2011)
	Bap1	Protein	Surface adhesion	(Berk, Fong et al. 2012)
Listeria monocytogenes	Teichoic acid	Polysaccha-ride	Surface adhesion	(Lin, Shu et al. 2015) (Berk, Fong et al. 2012, Brauge, Faille et al. 2018)
	Poly-β (1-6)-*N*-acetylglucosamine (PNAG)	Polysaccha-ride	Protective	(Köseoğlu, Heiss et al. 2015, Brauge, Faille et al. 2018)

In *Acinetobacter nosocomialis* and *Acinetobacter baumannii* (nosocomial pathogens that cause various human infections), the main outer membrane protein A (OmpA) plays a critical role in pathogenicity; is a key virulence factor that mediates biofilm formation, eukaryotic cell infection, antibiotic resistance, and immunomodulation. Deletion of *OmpA* gene resulted in a reduction of biofilm formation in abiotic surfaces (polystyrene tubes). Furthermore, reduced adherence to epithelial cells (A549) was observed compared to the wild-type strain. Adherence of these pathogenic bacteria to epithelial cells is an essential initial step for colonization and infection (Kim, Oh et al. 2016).

On the other hand, major components of the *V. cholerae* biofilm matrix are the exopolysaccharide VPS and the proteins RbmA, RbmC, and Bap1. It has been reported that RbmA is required for rugose colony morphology and the development of the *Vibrio* biofilm architecture. Previous evidence said that the absence of RbmA caused a loss of the strong structure, being *V. cholerae* biofilms more fragile and sensitive to detergent treatments (Giglio, Fong et al. 2013). Hahn, González et al. (2021) reported that *S.* Typhimurium mutations in the biofilm EPS, O antigen capsule, colonic acid and Vi antigen reduced the tolerance to oxidative stress caused by H_2O_2. This effect provides evidence of the important protective role of EPS in bacterial resistance to biocides. As mentioned previously, EPS components play a critical role in the adhesion of pathogenic bacteria to surfaces and biofilm formation. All these contribute to the remarkable resistance and persistence of pathogenic bacteria in different environments and the increased cases of infections. The secretion of EPS, the biofilm formation and other virulence determinants are controlled and regulated by intercellular communication and other molecular mechanisms in response to several stimuli such as environmental stresses like antimicrobials, desiccation, nutrient deficiency, among others.

Little is known about the biosynthetic and genetic mechanisms involved in the production of EPS components. One of these is colanic acid, and it is an extracellular heteropolysaccharide found in *E. coli*. Colanic acid comprises l-fucose, d-galactose, d-glucuronate, and d-glucose; these units are polymerized within the cytoplasm by glycosyltransferases excreted into the extracellular space by the Wzx system. Alginate, Psl, and Pel are found in EPS produced by *P. aeruginosa*. A study determined that the AlgC enzyme participates in the synthesis of alginate and Psl and Pel, and this enzyme, encoded by the *algC* gene (Rabin, Zheng et al. 2015). The polysaccharide Psl is accountable for promoting the initial binding process to the surface for the consequent formation of biofilms. The Pel genes are responsible for forming exopolysaccharides rich in glucose, and the *psl* genes are involved in the formation of exopolysaccharides rich in mannose in the *Pseudomonas* strains (Rabin, Zheng et al. 2015). There are 24 genes involved in the synthesis of EPS, eight genes that regulate the export of alginic acids, and 12 are responsible for biosynthesis, and four regulate the synthesis of polysaccharides (Rabin, Zheng et al. 2015) (Donot, Fontana et al. 2012). Therefore, a series of specific genes that work in a highly regulated manner are required to produce the proteins, which eventually perform the EPS synthesis (Nouha, Kumar et al. 2018).

On the other hand, *quorum sensing* plays an essential role in biofilm development. Pathogenic bacteria in biofilms use *quorum sensing* mechanisms to activate virulence and develop resistance to antibiotics. (Saxena, Joshi et al. 2019). Several biofilm bacteria release extracellular DNA (eDNA) that facilitates remodeling of the extracellular matrix and helps bind microbial cells into clumps by promoting acid-base interactions (Saxena, Joshi et al. 2019). While eDNA is commonly found in most biofilms, the externalization mechanisms of this DNA need further understanding. Recent reports suggest an autolysis-mediated and active secretion of eDNA in an exopolymeric matrix under the control of *quorum sensing*, eDNA serves a wide range of purposes in biofilm generation (Saxena, Joshi et al. 2019). It acts as a key adhesion molecule in the early stages of biofilm formation and as a structural component of mature bacterial aggregates and spasm motility structures. eDNA chelates and inactivates cationic antibiotics due to the inherent negative charge on the surface and contributes to antibiotic resistance. In some instances, eDNA has been shown to bind to Pel, the protein that facilitates cell-cell interactions within the biofilm and aids in resistance to antibiotics (Saxena, Joshi et al. 2019).

Consequently, biofilms represent a substantial challenge in clinical and industrial environments, and control methods are urgently required to prevent their development. The inhibition of bacteria EPS (synthesis or degradation) is a promising strategy to inhibit the cell adhesion affecting biofilm development. The inhibition of intracellular signaling and the inhibition of non-signaling mechanisms, which are involved in the secretion of the EPS, could indeed represent an effective strategy (Grande, Puca et al. 2020).

Strategies to Inhibit the Synthesis of Extracellular Polymeric Substances

Natural Sources

Nowadays, different products or compounds from natural origin have inhibited EPS synthesis that constitutes bacterial biofilms. In this sense, terpenes are the most explored group of compounds with this potential activity. Rubini, Banu et al. (2018) demonstrated that *Pogostemon heyneanus* and *Cinnamomum tomala* essential oils reduced the EPS of different strains of *S. aureus*, observed by scanning electron microscope analysis. Similarly, Cui, Zhang et al. (2020) analyzed the antibiofilm effect of cardamom essential oil (EO) against Methicillin-resistant *S. aureus*. Different EO concentrations (1/4MIC = 0.125, 1/2MIC = 0.25, MIC = 0.5 and 2MIC = 1 mg/mL) were tested in this study. The more active concentration observed against EPS production was 2MIC, which reduced the extracellular polysaccharide and protein by 61% and 58%, respectively. Additionally, there was no affectation on DNA content. The antibiofilm effect was associated with the regulation expression of the icaR gene and protein encoded by icaR, which could inhibit PIA synthesis.

Zhang, Li et al. (2020) studied the capacity of clove EO to affect *L. monocytogenes* biofilm formation. Results evidenced that all tested concentrations decrease extracellular polysaccharides' production (1/2 MBIC = 0.05, MBIC: 1.0, and 2MBIC = 2 mg/mL produce a reduction of ≈35%, ≈69%, ≈88%, respectively). Similar behavior was obtained in protein production due to a dose-dependence effect (1/2 MIC, MIC, and 2MIC produce a reduction of ≈34%, ≈44%, ≈76%, respectively) was observed. Also, findings demonstrated no effect on DNA content. Clove EO effect on biofilm formation could be related to the inhibition of *prfA* gene expression (84%), essential for biofilm formation. Yang, Rajivgandhi et al. (2020) obtained one fraction from *Hisbiscus rosa senenses* EO, affecting the EPS of *K. pneumoniae* biofilms. The obtained results showed that evaluated fractions exhibited a dose-dependence effect, getting exopolysaccharide inhibition values between 20 to 92% at doses between 20 to 100 μg/mL.

Weerasooriya reported similar findings, which demonstrated that *Ocimum tenuiflorum* extract showed the capacity to reduce the carbohydrate (56%), protein (22%) and DNA (44%) content of *S. aureus* biofilms. The previous effect was associated with the presence of majority compounds, eugenol, and linalool. The molecular docking analysis showed that these molecules could interact with a protein associated with biofilm formation. Similarly, Lahiri, Nag et al. (2021) performed an ethanolic extract from *Ocimim tenuiflorum* and eugenol and linalool. Results evidenced that at 175 μg/mL of ethanolic extract, 150 μg/mL of eugenol, and 175 μg/mL of linalool reduced carbohydrate concentration of *P. aeruginosa* biofilms, presenting 82%, 80%, and 70% less dose against the control group, respectively. These concentrations reduce the biofilm protein content by 85%, 82%, and 64% by ethanolic extract, eugenol, and linalool, respectively. The observed effect was associated with the capacity of evaluated treatment to inhibit the production of acyl-homoserine lactone, which plays an essential role in *P. aeruginosa* biofilm production.

Liu, Jin et al. (2021) demonstrated that carvacrol oil suppressed the production of the polysaccharide on *E. cloacae* biofilms, using 64 and 128 μg/mL concentrations with a subsequent 30% polysaccharide content reduction in biofilms at 24 h, 48 h, and 72 h, compared to the control. The effect was related to carvacrol's capacity to inhibit the expression of *csgABCEFG*, *fsrA*, *ftsQ* and *ftsZ* genes, which are important on *E. cloacae* biofilms production. Kannappan, Sivaranjani et al. (2017) analyzed the effect of geraniol on EPS of *S. epidermidis* biofilms. They reported that geraniol at 200 mg/mL and 150 mg/mL reduced 69% and 45% of the EPS production on *S. epidermidis* biofilms, respectively. In the same regard, Kannappan, Balasubramaniam et al. (2019) studied the antibiofilm effect of geraniol and cefotaxime combination against *Staphylococcus* spp. Results showed that the evaluated combination changed EPS from *S. epidermidis* and methicillin-resistant *S. aureus* (MRSA) biofilms, especially in proteins (amide I and amide II regions). Additionally, the evaluated treatment showed the capacity to inhibit the expression of genes superficial adhesion (*S. epidermidis*: *aap*, *sdrG*, *sdrF*, *bhp*, *ebh* and *altE*; MRSA: *fnbA*, *fnbB* and *clfA*), which may explain the capacity of the geraniol-cefotaxime combination to decrease the ability to produce biofilms.

Other compounds that have inhibited the EPS production in pathogenic bacteria biofilms are phenolic compounds. Information about this effect included research from Mooduto, Adittya et al. (2021), who observed that propolis extracts exhibited the capacity to affect the EPS production on *E. faecalis* biofilm, presenting 53% (0.2% of extract), 47% (0.8% of extract) and 28% (1.2% of extract) less EPS production after 24 h compared with control. The observed effect is associated with phenolic compounds in the extracts such as apigenin, tt-farnesol and tannins. Tao, Yan et al. (2021) analyzed the capacity of *Potentilla kleiniana* extracts to affect the EPS production in MRSA. Results evidenced that evaluated extract at 20 to 80 μg/mL decreased polysaccharides and protein formation from MRSA biofilms. The action mechanism is related to the *sarA*, *agrA* and *icaA* expression inhibition, important genes on *S. aureus* biofilm production.

Additionally, the identified compounds in extract were rutin, quercetin, naringin and 3-O-Methylducheside A. Vazquez-Armenta, Bernal-Mercado et al. (2018) demonstrated that quercetin at 0.2 mM decreased the production of the extracellular protein of *L. monocytogenes* biofilms, showing 41% reduction after 24 h compared with control (with no quercetin), while biofilm DNA content and polysaccharides were not affected by quercetin. In the same way, Sivaranjani, Srinivasan et al. (2018) observed that α-mangostin (2 μg/mL) strongly reduced polysaccharides (81% to 87%) and proteins (59% to 45%) in *A. baumannii* biofilms (two different bacterial strains). Additionally, the analysis by FT-IR showed that the evaluated treatment produced differences in EPS polysaccharides, proteins, and DNA. Changes in the intensity of peaks for glycosidic linkage in anomeric region and O-acetyl groups was also shown, all of them crucial for biofilm integrity. Thus, authors associated this effect with the downregulation on *pgaC* and *vpgaA* gene expression, which are involved in the initial stages from *A. baumannii* biofilm formation.

Alkaloids are another group of secondary metabolites that exhibited an effect on EPS from pathogenic bacterial biofilms. In addition, it has been observed that berberine produced a protein and eDNA to reduce the EPS from *P. aeruginosa* biofilms, used at concentrations of 512, 1024, and 2048 μg/mL. This effect was also demonstrated by microscopy SEM, where biofilm structure changes were evidenced. Additionally, the compound decreased gene expression (*lasl, rhll, rhlR,* and *lasR*) related to *P. aeruginosa* biofilm formation (Li, Huang et al. 2017). In the same way, caffeine exhibited the ability to decrease EPS (40% and 50%, respectively) and protein (20% and 25%, respectively) production from *P. aeruginosa* biofilms at doses of 40 and 80 μg/mL. Additionally, caffeine showed the capacity to alter the motility and *quorum sensing* proteins (LasR and LasI), affecting bacterial biofilm formation (Chakraborty, Dastidar et al. 2020).

Aldehydes have shown the ability to inhibit EPS production in pathogenic bacteria biofilms. Vijayakumar and Thirunanasambandham (2021) tested the antibiofilm effect of 5-hydroxymethylfurfural against *A. baumannii*. Results evidenced that the evaluated aldehyde (100 μg/mL) produced an important diminution in polysaccharides and protein in *A. baumannii* biofilms, 85% and 74% less than the control, respectively. Additionally, the expression inhibition of *bap, bfmR, csuA/B, ompA,* and *katE* was also observed, which are involved in the adhesion process and the subsequent biofilm formation.

Another type of compound that has presented antibiofilm effects is anthraquinones. Yan, Gu et al. (2017) demonstrated that emodin showed a dose-dependence inhibition effect on eDNA EPS production from *S. aureus* biofilms, presenting an eDNA reduction between 0 to 83%, the 4 MIC = 256 µg/mL was the concentration most effective. Also, the SEM analysis confirmed that the dose of ¼ MIC = 16 µg/mL and ½ MIC = 32 µg/mL significantly affected EPS biofilms, causing an alteration in their structure. Additionally, it was reported that emodin caused downregulation of *dltB, SarA, SortaseA, agrA, icaA,* and *cidA* genes, which are implicated in *S. aureus* biofilm formation.

Enzymatic Treatments

Other alternatives that demonstrated being highly able to affect EPS production from pathogenic bacteria biofilms are enzymatic treatments. In this regard, Nahar, Ha et al. (2021) evaluated the efficacy of flavourzyme to affect EPS production of different pathogens. Initially, FTIR analyzed that flavourzyme induced changes in bans corresponding to EPS proteins and carbohydrates from *S. Typhimurium, E. coli,* and *P. aeruginosa* at the different evaluated doses (200 to 800 µL/mL). Additionally, it was observed that total protein content from the EPS biofilm of the three pathogens was eliminated at least 95% at the dose of 800 µL/mL in both analyzed surfaces (ultra-high molecular weight polyethylene and rubber). Also, the same tested concentration (800 µL/mL) exhibited the highest capacity to reduce EPS biofilm carbohydrates content (30 to 97%) of evaluated pathogens in the analyzed surfaces, being *P. aeruginosa* the most susceptible, followed by *E. coli* and *S.* Typhimurium. Also, field emission scanning electron microscopy demonstrated that all evaluated concentrations of flavourzyme in both studied surfaces produced a disruption of the EPS matrix on all evaluated pathogens, being the concentration of 800 µL/mL the most active treatment.

Similarly, other author observed that α-amylase exhibited the capacity to disrupt the EPS structure from different human pathogens biofilms, affecting the content of carbohydrates and protein from MRSA (carbohydrates reduction: 31 to 43%; protein reduction: 3 to 31%) *V. cholerae* (carbohydrates reduction: 37 to 64%; protein reduction: 23 to 24%) and *P. aeruginosa* (carbohydrates reduction: 23 to 42%; protein reduction: 4 to 18%) biofilms. Added to this, throughout *in-silico* analysis was possible to evidenced that enzyme can interact by a spontaneous process with the EPS components (carbohydrates and proteins), generating high free energy values, explaining the inhibitory effect on the biofilm formation (Kalpana, Aarthy et al. 2012, Lahiri, Nag et al. 2021).

Similarly, Molobela, Cloete et al. (2010) reported that proteases savinase and everlase effectively reduced EPS protein content from *P. fluorescens* biofilms. In addition, amyloglucosidase and amylase significantly decreased EPS carbohydrate content from *P. fluorescens* biofilms. Scanning electron microscopy assay demonstrated that the evaluated enzymes induced damage in biofilm integrity. Research performed by Nagraj and Gokhale (2018) assessed the enzyme complex efficacy of amylase, cellulase, and protease (Maximum specific enzyme activities of 3.04, 2.61, and 3.39 IU/mg, respectively) affect EPS from *E. coli*, *Salmonella enterica*, *P. aeruginosa*, and *S. aureus* in an 85, 79, 88, 87%, respectively. Other enzymes such as DNase I (1 mg/mL) effectively reduced eDNA content (more than 90%) of *V. parahaemolyticus* biofilms. Additionally, alteration in biofilm structure was observed by SEM and TEM assays. Findings indicated that enzymes effectively inhibited the expression of essential genes (*aphA, opaR, cpsA, cpsQ* and *cpsR*) involved in the *V. parahaemolyticus* biofilm formation.

Chemical Synthetic Treatments

Das, Paul et al. (2016) demonstrated that different 3-amino-4-aminoximidofurazan derivatives were effective in decreasing EPS content from *S. aureus* (at least 70%) and *P. aeruginosa* biofilms (at least 75%) in the range of 60-170 μg/mL. Also, a diminution in protein content from *S. aureus* (at least 30%) and *P. aeruginosa* biofilms (at least 50%) was obtained. The stained with acridine orange revealed changes in EPS biofilm structure. These results were associated with the chemical compounds' capacity to inhibit the motility assay, an important process associated with biofilm formation. Bhattacharyya, Agarwal et al. (2018) reported that zinc oxide was able to decrease the exopolysaccharide content from *S. pneumoniae* biofilms in a dose-dependence relation, being the concentration of 12 μg/mL the most effective (≈35% of reduction). In a study performed by Li, Tan et al. (2020), acidic electrolytic water affected the EPS from *V. parahaemolyticus* biofilms. The evaluated treatment (1 mg/mL) reduced 90%, 30%, and 50% in eDNA, total carbohydrates, and total protein from *V. parahaemolyticus* biofilms. Additionally, the SEM and TEM analysis revealed that acidic electrolytic water (1 mg/mL) produced critical changes in biofilm structure. Also, this treatment produced an inhibition of the *aphA, opaR, cpsA, cpsQ* and *cpsR* expression, which are involved in the biofilm formation process.

Nanoparticles

Nanoparticles inhibited the EPS biofilm formation from pathogenic bacteria. Rajivgandhi, Kanisha et al. (2021) reported that silver oxide nanoparticles could reduce EPS production from *K. pneumonia* biofilms, showing inhibition values between 20 to 85% at different tested concentrations (10 to 200 μg/mL). For all cases, high concentration promoted high activity. Additionally, the alteration in biofilm extracellular structure (SEM study) was observed. Similarly, zinc oxide nanoparticles showed potential affecting EPS production (higher than 90%) from MRSA biofilms, and confocal microscopy analysis confirmed this effect. Additionally, the mechanism of actions is related to nanoparticles' capacity to affect the motility and slime process and induced reactive oxygen species production (Banerjee, Vishakha et al. 2020).

On the other hand, nanoparticles have been combined with other antimicrobial strategies (plant extracts) to affect the bacterial biofilm formation process. Vanaraj, Keerthana et al. (2017) tested the ability of silver nanoparticles incorporated with quercetin from *Clitotia ternatea* to decrease EPS from *S. aureus*. Authors observed that concentrations of 25 μg/mL, 50 μg/mL, 75 μg/mL, and 100 μg/mL inhibited the EPS productions being dependent on doses, obtained around ≈18%, ≈30%,

≈60%, and ≈75% of inhibition, respectively. Similarly, Ravindran, Ramanathan et al. (2018) reported that silver nanoparticles incorporate *Vetiveria zizanoides* extract showed the capacity to reduce EPS production from *S. marcescens* biofilms, exhibiting 8%, 28%, and 60% inhibition at 0.5, 1, and 2 µg/mL, respectively. An alteration of biofilms structure by microscopic assay was also observed. Nanoparticles were able to inhibit the gene expression (*fimA, fimC, flhD* and *bsmB*) involved in the *quorum sensing*, which may explain the antibiofilm effect.

In the same regard, silver nanoparticles combined with *Gracilaria corticata* affected the production of EPS *K. pneumoniae* biofilms, showing inhibition values between 20% to 80% (concentration: 20 to 80 µg/mL) the highest concentration was the most effective. The SEM study also demonstrated this effect, which was evident in the alteration in standard biofilm architecture (Rajivgandhi, Ramachandran et al. 2020). Gold nanoparticles incorporate with *Capsicum annuum* extract also demonstrated efficacy to reduce the EPS production of *P. aeruginosa* and *S. marcescens* at 25 (≈15 and ≈10%, respectively), 50 (≈30 and ≈28%, respectively), 100 (≈60 and ≈55%, respectively) and 200 µg/mL (≈85 and ≈75%, respectively). The microscopy analysis corroborated that all evaluated concentrations induced biofilm structure changes (Qais, Ahmad et al. 2021).

Conclusion

The study of EPS-producing pathogenic bacteria is an area that should be further explored as it is a field of study with a wealth of research opportunities for many industrial and ecological applications. Biofilms of pathogenic bacteria largely depend on the type of strain, availability of nutrients and growing conditions. For some bacteria, it has been possible to identify the sequence of genes that encode the production of EPS, composition, and synthesis's metabolic route. However, many pathogenic bacteria cause specific diseases where it has not been possible to establish a relationship between secretory mechanisms, genes, and EPS production, together with enzymes involved in synthesis and excretion. EPS production can already be controlled in numerous cases; however, the antibiotic dose used is much higher in biofilm. For this reason, molecular analyses can lead to a greater understanding of the subject and unravel the molecular mechanisms that underlie the existence of the microbial community. This knowledge will allow us to establish more specific strategies for eliminating and eradicating biofilms, notably reducing the cases of infectious diseases caused by bacteria products of EPS.

References

Adator, E. H., M. Cheng, R. Holley, T. McAllister and C. Narvaez-Bravo (2018). "Ability of Shiga toxigenic *Escherichia coli* to survive within dry-surface biofilms and transfer to fresh lettuce." *International Journal of Food Microbiology* 269: 52-59.

AlEraky, D. M., M. Madi, M. El Tantawi, J. AlHumaid, S. Fita, S. AbdulAzeez, J. F. Borgio, F. A. Al-Harbi and A. S. Alagl (2021). "Predominance of non-*Streptococcus mutans* bacteria in dental biofilm and its relation to caries progression." *Saudi Journal of Biological Sciences*. In-press.

Amrutha, B., K. Sundar and P. H. Shetty (2017). "Effect of organic acids on biofilm formation and quorum signaling of pathogens from fresh fruits and vegetables." *Microbial Pathogenesis* 111: 156-162.

Aryal, M. and P. M. Muriana (2019). "Efficacy of Commercial Sanitizers Used in Food Processing Facilities for Inactivation of *Listeria monocytogenes*, *E. coli* O157:H7, and *Salmonella* Biofilms." *Foods* 8(12): 639.

Ashrafudoulla, M., K. W. Na, M. I. Hossain, M. F. R. Mizan, S. Nahar, S. H. Toushik, P. K. Roy, S. H. Park and S.-D. J. M. P. B. Ha (2021). "Molecular and pathogenic characterization of *Vibrio parahaemolyticus* isolated from seafood." *Marine Pollution Bulletin* 172: 112927.

Bales, P. M., E. M. Renke, S. L. May, Y. Shen and D. C. Nelson (2013). "Purification and Characterization of Biofilm-Associated EPS Exopolysaccharides from ESKAPE Organisms and Other Pathogens." *PLOS ONE* 8(6): e67950.

Balière, C., A. Rincé, J. Blanco, G. Dahbi, J. Harel, P. Vogeleer, J.-C. Giard, P. Mariani-Kurkdjian and M. Gourmelon (2015). "Prevalence and characterization of Shiga toxin-producing and enteropathogenic *Escherichia coli* in shellfish-harvesting areas and their watersheds." *Frontiers in Microbiology* 6: 1356.

Banerjee, S., K. Vishakha, S. Das, M. Dutta, D. Mukherjee, J. Mondal, S. Mondal and A. Ganguli (2020). "Antibacterial, anti-biofilm activity and mechanism of action of pancreatin doped zinc oxide nanoparticles against methicillin resistant *Staphylococcus aureus*." *Colloids and Surfaces B: Biointerfaces* 190: 110921.

Behzadi, P., E. Urbán and M. J. D. Gajdács (2020). "Association between biofilm-production and antibiotic resistance in uropathogenic *Escherichia coli* (UPEC): an *in vitro* study." *Diseases* 8(2): 17.

Belfield, K., R. Bayston, J. P. Birchall and M. Daniel (2015). "Do orally administered antibiotics reach concentrations in the middle ear sufficient to eradicate planktonic and biofilm bacteria? A review." *International Journal of Pediatric Otorhinolaryngology* 79(3): 296-300.

Beloin, C., A. Roux and J.-M. Ghigo (2008). *Escherichia coli* Biofilms. *Bacterial Biofilms*. T. Romeo. Berlin, Heidelberg, Springer Berlin Heidelberg: 249-289.

Berk, V., J. C. N. Fong, G. T. Dempsey, O. N. Develioglu, X. Zhuang, J. Liphardt, F. H. Yildiz and S. Chu (2012). "Molecular architecture and assembly principles of *Vibrio cholerae* biofilms." *Science (New York, N.Y.)* 337(6091): 236-239.

Bernal-Mercado, A. T., M. M. Gutierrez-Pacheco, D. Encinas-Basurto, V. Mata-Haro, A. A. Lopez-Zavala, M. A. Islas-Osuna, G. A. Gonzalez-Aguilar and J. F. Ayala-Zavala (2020). "Synergistic mode of action of catechin, vanillic and protocatechuic acids to inhibit the adhesion of uropathogenic *Escherichia coli* on silicone surfaces." *Journal of Applied Microbiology* 128(2): 387-400.

Bhardwaj, D. K., N. K. Taneja, S. Dp, A. Chakotiya, P. Patel, P. Taneja, D. Sachdev, S. Gupta and M. G. Sanal (2021). "Phenotypic and genotypic characterization of biofilm forming, antimicrobial resistant, pathogenic *Escherichia coli* isolated from Indian dairy and meat products." *International Journal of Food Microbiology* 336: 108899.

Bhattacharyya, P., B. Agarwal, M. Goswami, D. Maiti, S. Baruah and P. Tribedi (2018). "Zinc oxide nanoparticle inhibits the biofilm formation of *Streptococcus pneumoniae*." *Antonie Van Leeuwenhoek* 111(1): 89-99.

Billings, N., M. Ramirez Millan, M. Caldara, R. Rusconi, Y. Tarasova, R. Stocker and K. Ribbeck (2013). "The Extracellular Matrix Component Psl Provides Fast-Acting Antibiotic Defense in *Pseudomonas aeruginosa* Biofilms." *PLOS Pathogens* 9(8): e1003526.

Boswell, M. T. and R. J. S. A. J. o. I. D. Cockeran (2021). "Effect of antimicrobial peptides on planktonic growth, biofilm formation and biofilm-derived bacterial viability of *Streptococcus pneumoniae*." *Southern African Journal of Infectious Diseases* 36(1): 226.

Brauge, T., C. Faille, I. Sadovskaya, A. Charbit, T. Benezech, Y. Shen, M. J. Loessner, J. R. Bautista and G. Midelet-Bourdin (2018). "The absence of N-acetylglucosamine in wall teichoic acids of *Listeria monocytogenes* modifies biofilm architecture and tolerance to rinsing and cleaning procedures." *PLOS ONE* 13(1): e0190879.

Brindhadevi, K., F. LewisOscar, E. Mylonakis, S. Shanmugam, T. N. Verma and A. J. P. B. Pugazhendhi (2020). "Biofilm and Quorum sensing mediated pathogenicity in *Pseudomonas aeruginosa*." *Process Biochemistry* 96: 49-57.

Brown, H. L., M. Reuter, L. J. Salt, K. L. Cross, R. P. Betts, A. H. M. v. Vliet and M. W. Griffiths (2014). "Chicken Juice Enhances Surface Attachment and Biofilm Formation of *Campylobacter jejuni*." *Applied and Environmental Microbiology* 80(22): 7053-7060.

Byun, K.-H., S. H. Han, J.-w. Yoon, S. H. Park and S.-D. J. F. C. Ha (2021). "Efficacy of chlorine-based disinfectants (sodium hypochlorite and chlorine dioxide) on *Salmonella Enteritidis* planktonic cells, biofilms on food contact surfaces and chicken skin." *Food Control* 123: 107838.

CDC. (2020). *Outbreak of E. coli Infections Linked to Leafy Greens*. Retrieved September 24, 2021, from https://www.cdc.gov/ecoli/2020/o157h7-10-20b/index.html.

CDC. (2021). *List of Selected Multistate Foodborne Outbreak Investigations*. Retrieved September 22, 2021.

CDC. (2021). *Reports of Selected Salmonella Outbreak Investigations*. Retrieved September 24, 2021, from https://www.cdc.gov/salmonella/outbreaks.html.

CDC. (2021). *Salmonella Outbreak Linked to Frozen Cooked Shrimp*. Retrieved September 24, 2021, from https://www.cdc.gov/salmonella/weltevreden-06-21/index.html.

CDC, C. f. D. C. a. P. (2021, July 15, 2021). *Foodborne Outbreaks*. Retrieved July 17, 2021, from https://www.cdc.gov/foodsafety/outbreaks/multistate-outbreaks/outbreaks-list.html.

Cerca, N., T. Maira-Litrán, K. K. Jefferson, M. Grout, D. A. Goldmann and G. B. Pier (2007). "Protection against *Escherichia coli* infection by antibody to the *Staphylococcus aureus* poly-N-acetylglucosamine surface polysaccharide." *Proceedings of the National Academy of Sciences* 104(18): 7528-7533.

Chakraborty, P., D. G. Dastidar, P. Paul, S. Dutta, D. Basu, S. R. Sharma, S. Basu, R. K. Sarker, A. Sen and A. Sarkar (2020). "Inhibition of biofilm formation of *Pseudomonas aeruginosa* by caffeine: A potential approach for sustainable management of biofilm." *Archives of Microbiology* 202(3): 623-635.

Chen, Y., W. H. Ross, R. C. Whiting, A. Van Stelten, K. K. Nightingale, M. Wiedmann and V. N. Scott (2011). "Variation in *Listeria monocytogenes* dose responses in relation to subtypes encoding a full-length or truncated internalin A." *Applied and Environmental Microbiology* 77(4): 1171-1180.

Choi, A. H., L. Slamti, F. Y. Avci, G. B. Pier and T. Maira-Litrán (2009). "The pgaABCD locus of *Acinetobacter baumannii* encodes the production of poly-β-1-6-N-acetylglucosamine, which is critical for biofilm formation." *Journal of Bacteriology* 191(19): 5953-5963.

Colwell, R. R., A. Huq, M. S. Islam, K. Aziz, M. Yunus, N. H. Khan, A. Mahmud, R. B. Sack, G. B. Nair and J. J. P. o. t. N. A. o. S. Chakraborty (2003). "Reduction of cholera in Bangladeshi villages by simple filtration." *Proceedings of the National Academy of Sciences of the United States of America* 100(3): 1051-1055.

Cui, H., C. Zhang, C. Li and L. Lin (2020). "Inhibition mechanism of cardamom essential oil on methicillin-resistant *Staphylococcus aureus* biofilm." *LWT* 122: 109057.

Das, A. K., P. K. Nanda, A. Das and S. Biswas (2019). Chapter 6 - Hazards and Safety Issues of Meat and Meat Products. *Food Safety and Human Health*. R. L. Singh and S. Mondal, Academic Press: 145-168.

Das, M., S. Paul, P. Gupta, P. Tribedi, S. Sarkar, D. Manna and S. Bhattacharjee (2016). "3-Amino-4-aminoximidofurazan derivatives: small molecules possessing antimicrobial and antibiofilm activity against *Staphylococcus aureus* and *Pseudomonas aeruginosa*." *Journal of Applied Microbiology* 120(4): 842-859.

Dincer, S., M. S. Özdenefe and A. Arkut (2020). *Bacterial Biofilms*, IntechOpen.

Domenech, M., E. García and M. J. M. b. Moscoso (2012). "Biofilm formation in *Streptococcus pneumoniae*." *Microbial Biotechnology* 5(4): 455-465.

Donot, F., A. Fontana, J. Baccou and S. J. C. P. Schorr-Galindo (2012). "Microbial exopolysaccharides: main examples of synthesis, excretion, genetics and extraction." *Carbohydrate Polymers* 87(2): 951-962.

Faille, C., C. Cunault, T. Dubois and T. Benezech (2018). "Hygienic design of food processing lines to mitigate the risk of bacterial food contamination with respect to environmental concerns." *Innovative Food Science & Emerging Technologies* 46: 65-73.

Flemming, H.-C., J. Wingender, U. Szewzyk, P. Steinberg, S. A. Rice and S. Kjelleberg (2016). "Biofilms: an emergent form of bacterial life." *Nature Reviews Microbiology* 14(9): 563-575.

Flemming, H.-C. and J. J. N. r. m. Wingender (2010). "*The biofilm matrix.*" 8(9): 623-633.

Galie, S., C. García-Gutiérrez, E. M. Miguélez, C. J. Villar and F. Lombó (2018). "Biofilms in the food industry: health aspects and control methods." *Frontiers in Microbiology* 9: 898.

Gao, T., Y. Ding, Q. Wu, J. Wang, J. Zhang, S. Yu, P. Yu, C. Liu, L. Kong and Z. Feng (2018). "Prevalence, virulence genes, antimicrobial susceptibility, and genetic diversity of *Bacillus cereus* isolated from pasteurized milk in China." *Frontiers in Microbiology* 9: 533.

Giglio, K. M., J. C. Fong, F. H. Yildiz and H. Sondermann (2013). "Structural basis for biofilm formation via the *Vibrio cholerae* matrix protein RbmA." *Journal of Bacteriology* 195(14): 3277-3286.

Gil, M. I., M. V. Selma, T. Suslow, L. Jacxsens, M. Uyttendaele, and Ana Allende (2015). "Pre-and postharvest preventive measures and intervention strategies to control microbial food safety hazards of fresh leafy vegetables." *Critical Reviews in Food Science and Nutrition* 55(4): 453-468.

Gopal, N., C. Hill, P. R. Ross, T. P. Beresford, M. A. Fenelon and P. D. Cotter (2015). "The Prevalence and Control of *Bacillus* and Related Spore-Forming Bacteria in the Dairy Industry." *Frontiers in Microbiology* 6(1418).

Grande, R., V. Puca and R. Muraro (2020). *Antibiotic resistance and bacterial biofilm*, Taylor & Francis. 30 (12): 897-900.

Griffin, P. M. and M. A. Karmali (2016). "Emerging Public Health Challenges of Shiga Toxin–Producing *Escherichia coli* Related to Changes in the Pathogen, the Population, and the Environment." *Clinical Infectious Diseases* 64(3): 371-376.

Gupta, P., S. Sarkar, B. Das, S. Bhattacharjee and P. Tribedi (2016). "Biofilm, pathogenesis and prevention—a journey to break the wall: a review." *Archives of Microbiology* 198(1): 1-15.

Gupta, P., S. Sarkar, B. Das, S. Bhattacharjee and & Tribedi, P. (2016). "Biofilm, pathogenesis and prevention—a journey to break the wall: a review." *Archives in Microbiology* 198(1): 1-15.

Gutiérrez-Pacheco, M. M., A. T. Bernal-Mercado, F. J. Vázquez-Armenta, M. A. Martínez-Tellez, G. A. González-Aguilar, J. Lizardi-Mendoza, T. J. Madera-Santana, F. Nazzaro and J. F. Ayala-Zavala (2019). "*Quorum sensing* interruption as a tool to control virulence of plant pathogenic bacteria." *Physiological and Molecular Plant Pathology* 106: 281-291.

Hahn, M. M., J. F. González and J. S. Gunn (2021). "*Salmonella* Biofilms Tolerate Hydrogen Peroxide by a Combination of Extracellular Polymeric Substance Barrier Function and Catalase Enzymes." *Frontiers in Cellular and Infection Microbiology* 11(432).

Hameed, Z. R., A. S. Motib, A. F. Abbas (2021). "Adaptability of Biofilm Formation in *Streptococcus pneumoniae* to Various Growth Conditions" *Indian Journal of Forensic Medicine & Toxicology* 15(2).

Han, R., Y. A. K. Klu and J. Chen (2017). "Attachment and biofilm formation by selected strains of *Salmonella enterica* and entrohemorrhagic *Escherichia coli* of fresh produce origin." *Journal of Food Science* 82(6): 1461-1466.

Hancock, R. E., M. A. Alford and E. F. Haney (2021). "Antibiofilm activity of host defence peptides: complexity provides opportunities." *Nature Reviews Microbiology* 1-12.

Henriques, A. R., L. Telo da Gama and M. J. Fraqueza (2014). "Assessing *Listeria monocytogenes* presence in Portuguese ready-to-eat meat processing industries based on hygienic and safety audit." *Food Research International* 63: 81-88.

Hingston, P. A., E. C. Stea, S. Knøchel and T. Hansen (2013). "Role of initial contamination levels, biofilm maturity and presence of salt and fat on desiccation survival of *Listeria monocytogenes* on stainless steel surfaces." *Food Microbiology* 36(1): 46-56.

Huang, Y., S. H. Flint and J. S. Palmer (2020). "*Bacillus cereus* spores and toxins–The potential role of biofilms." *Food Microbiology* 90: 103493.

Huffnagle, G., R. Dickson and N. J. M. i. Lukacs (2017). "The respiratory tract microbiome and lung inflammation: a two-way street." *Mucosal Immunology* 10(2): 299-306.

Izadi, P., P. Izadi and A. J. C. Eldyasti (2021). "Holistic insights into extracellular polymeric substance (EPS) in anammosx bacterial matrix and the potential sustainable biopolymer recovery: A review." *Chemosphere* 129703.

Jennings, L. K., K. M. Storek, H. E. Ledvina, C. Coulon, L. S. Marmont, I. Sadovskaya, P. R. Secor, B. S. Tseng, M. Scian, A. Filloux, D. J. Wozniak, P. L. Howell and M. R. Parsek (2015). "Pel is a cationic exopolysaccharide that cross-links extracellular DNA in the *Pseudomonas aeruginosa* biofilm matrix." *Proceedings of the National Academy of Sciences*, 112(36): 11353-11358.

Joshi, K., R. Mahendran, K. Alagusundaram, T. Norton, B. K. (2013). "Novel disinfectants for fresh produce." *Trends in Food Science & Technology* 34(1): 54-61.

Kalpana, B. J., S. Aarthy and S. K. Pandian (2012). "Antibiofilm activity of α-amylase from *Bacillus subtilis* S8-18 against biofilm forming human bacterial pathogens." *Applied Biochemistry and Biotechnology* 167(6): 1778-1794.

Kannappan, A., B. Balasubramaniam, B. Ranjitha, R. Srinivasan, R. Packiavathy, I. A. S. V. Balamurugan, S. K. Pandian and A. V. Ravi (2019). "*In vitro* and *in vivo* biofilm inhibitory efficacy of geraniol-cefotaxime combination against *Staphylococcus spp*." *Food and Chemical Toxicology* 125: 322-332.

Kannappan, A., M. Sivaranjani, R. Srinivasan, J. Rathna, S. K. Pandian and A. V. Ravi (2017). "Inhibitory efficacy of geraniol on biofilm formation and development of adaptive resistance in *Staphylococcus epidermidis* RP62A." *Journal of Medical Microbiology* 66(10): 1506-1515.

Karki, A. B., K. Ballard, C. Harper, R. J. Sheaff and M. K. J. S. r. Fakhr (2021). "*Staphylococcus aureus* enhances biofilm formation, aerotolerance, and survival of *Campylobacter* strains isolated from retail meats." *Scientific Reports* 11(1): 1-10.

Karygianni, L., Z. Ren, H. Koo and T. Thurnheer (2020). "Biofilm Matrixome: Extracellular Components in Structured Microbial Communities." *Trends in Microbiology* 28(8): 668-681.

Khanashyam, A. C., M. A. Shanker, A. Kothakota, N. K. Mahanti, R. J. O. S. Pandiselvam and Engineering (2021). "Ozone Applications in Milk and Meat Industry." *Science & Engineering* 1-16.

Khelissa, S. O., M. Abdallah, C. Jama, C. Faille and N.-E. Chihib (2017). "Bacterial contamination and biofilm formation on abiotic surfaces and strategies to overcome their persistence." *Journal of Materials and Environmental Science* 8(9): 3326-3346.

Kim, S. W., M. H. Oh, S. H. Jun, H. Jeon, S. I. Kim, K. Kim, Y. C. Lee and J. C. Lee (2016). "Outer membrane Protein A plays a role in pathogenesis of *Acinetobacter nosocomialis*." *Virulence* 7(4): 413-426.

Köseoğlu, V. K., C. Heiss, P. Azadi, E. Topchiy, Z. T. Güvener, T. E. Lehmann, K. W. Miller and M. Gomelsky (2015). "*Listeria monocytogenes* exopolysaccharide: origin, structure, biosynthetic machinery and c-di-GMP-dependent regulation." *Molecular Microbiology* 96(4): 728-743.

Kumar, S., N. Chandra, L. Singh, M. Z. Hashmi and A. Varma (2019). *Biofilms in Human Diseases: Treatment and Control*, Springer.

Kyere, E. O., G. Foong, J. Palmer, J. J. Wargent, G. C. Fletcher and S. Flint (2020). "Biofilm formation of *Listeria monocytogenes* in hydroponic and soil grown lettuce leaf extracts on stainless steel coupons." *LWT* 126: 109114.

Lahiri, D., M. Nag, B. Dutta, S. Dey, D. Mukherjee, S. J. Joshi and R. R. Ray (2021). "Antibiofilm and Anti-*Quorum sensing* Activities of Eugenol and Linalool from *Ocimum tenuiflorum* against *Pseudomonas aeruginosa* Biofilm." *Journal of Applied Microbiology.*

Lahiri, D., M. Nag, T. Sarkar, B. Dutta and R. R. Ray (2021). "Antibiofilm activity of α-Amylase from *Bacillus subtilis* and prediction of the optimized conditions for biofilm removal by response surface methodology (RSM) and Artificial Neural Network (ANN)." *Applied Biochemistry and Biotechnology* 193(6): 1853-1872.

Li, J., J. Feng, L. Ma, C. de la Fuente Núñez, G. Gölz and X. Lu (2017). "Effects of meat juice on biofilm formation of *Campylobacter* and *Salmonella*." *International Journal of Food Microbiology* 253: 20-28.

Li, Y., J. Huang, L. Li and L. Liu (2017). "Synergistic activity of berberine with azithromycin against *Pseudomonas aeruginosa* isolated from patients with cystic fibrosis of lung *in vitro* and *in vivo*." *Cellular Physiology and Biochemistry* 42(4): 1657-1669.

Li, Y., L. Tan, L. Guo, P. Zhang, P. K. Malakar, F. Ahmed, H. Liu, J. J. Wang and Y. Zhao (2020). "Acidic electrolyzed water more effectively breaks down mature *Vibrio parahaemolyticus* biofilm than DNase I." *Food Control* 117: 107312.

Liang, Z. X. (2015). "The expanding roles of c-di-GMP in the biosynthesis of exopolysaccharides and secondary metabolites." *Natural Product Reports* 32(5): 663-683.

Limoli, D. H., C. J. Jones, D. J. Wozniak, M. Ghannoum, M. Parsek, M. Whiteley and P. Mukherjee (2015). "Bacterial Extracellular Polysaccharides in Biofilm Formation and Function." *Microbiology Spectrum* 3(3): 3.3.29.

Lin, M. H., J. C. Shu, L. P. Lin, K. y. Chong, Y. W. Cheng, J. F. Du and S.-T. Liu (2015). "Elucidating the Crucial Role of Poly N-Acetylglucosamine from *Staphylococcus aureus* in Cellular Adhesion and Pathogenesis." *PLOS ONE* 10(4): e0124216.

Liu, F., P. Jin, Z. Sun, L. Du, D. Wang, T. Zhao and M. P. Doyle (2021). "Carvacrol oil inhibits biofilm formation and exopolysaccharide production of *Enterobacter cloacae*." *Food Control* 119: 107473.

Luke, N. R., J. A. Jurcisek, L. O. Bakaletz, A. A. J. I. Campagnari (2007). "Contribution of *Moraxella catarrhalis* type IV pili to nasopharyngeal colonization and biofilm formation." *Infection and Immunity* 75(12): 5559-5564.

Macarisin, D., J. Patel, G. Bauchan, J. A. Giron and V. K. Sharma (2012). "Role of Curli and Cellulose Expression in Adherence of *Escherichia coli* O157:H7 to Spinach Leaves." *Foodborne Pathogens and Disease* 9(2): 160-167.

Magennis, E. P., N. Francini, F. Mastrotto, R. Catania, M. Redhead, F. Fernandez-Trillo, D. Bradshaw, D. Churchley, K. Winzer and C. J. P. o. Alexander (2017). "Polymers for binding of the gram-positive oral pathogen *Streptococcus mutans*." *PloS one* 12(7): e0180087.

Manios, S. G., A. E. Kapetanakou, E. Zilelidou, S. Poimenidou and P. N. Skandamis (2014). Mechanisms and Risks Associated with Bacterial Transfer between Abiotic and Biotic Surfaces. *Microbial Food Safety and Preservation Techniques*, CRC Press: 108-133.

Marques, V. F., C. C. d. Motta, B. d. S. Soares, D. A. d. Melo, S. d. M. d. O. Coelho, I. d. S. Coelho, H. S. Barbosa and M. M. S. d. Souza (2017). "Biofilm production and beta-lactamic resistance in Brazilian *Staphylococcus aureus* isolates from bovine mastitis." Brazilian *Journal of Microbiology* 48: 118-124.

Maruzani, R., G. Sutton, P. Nocerino and M. Marvasi (2019). "Exopolymeric substances (EPS) from *Salmonella enterica*: polymers, proteins and their interactions with plants and abiotic surfaces." *Journal of Microbiology* 57(1): 1-8.

Meng, L., Y. Zhang, H. Liu, S. Zhao, J. Wang and N. Zheng (2017). "Characterization of *Pseudomonas spp.* and Associated Proteolytic Properties in Raw Milk Stored at Low Temperatures." *Frontiers in Microbiology* 8(2158).

Mizan, M. F. R., I. K. Jahid and S.-D. Ha (2015). "Microbial biofilms in seafood: A food-hygiene challenge." *Food Microbiology* 49: 41-55.

Molobela, I. P., T. E. Cloete and M. Beukes (2010). "Protease and amylase enzymes for biofilm removal and degradation of extracellular polymeric substances (EPS) produced by *Pseudomonas fluorescens* bacteria." *African Journal of Microbiology Research* 4(14).

Mooduto, L., D. Adittya, A. Subiyanto, A. Bhardwaj, Z. Arwidhyan, S. Goenharto and D. A. Wahjuningrum (2021). "The Effectiveness of Propolis Extract against Extracellular Polymeric Substance (EPS) Biofilm *Enterococcus Faecalis* Bacteria." *Journal of International Dental and Medical Research* 14(1): 54-59.

Mritunjay, S. K. and V. Kumar (2015). "Fresh farm produce as a source of pathogens: a review." *Research Journal of Environmental Toxicology* 9(2): 59-70.

Muhterem-Uyar, M., M. Dalmasso, A. S. Bolocan, M. Hernandez, A. E. Kapetanakou, T. Kuchta, S. G. Manios, B. Melero, J. Minarovičová, A. I. Nicolau, J. Rovira, P. N. Skandamis, K. Jordan, D. Rodríguez-

Lázaro, B. Stessl and M. Wagner (2015). "Environmental sampling for *Listeria monocytogenes* control in food processing facilities reveals three contamination scenarios." *Food Control* 51: 94-107.

Mulcahy, L. R., V. M. Isabella and K. Lewis (2014). "*Pseudomonas aeruginosa* biofilms in disease." *Microbial Ecology* 68(1): 1-12.

Murdoch, D. R. and S. R. Howie (2018). "The global burden of lower respiratory infections: making progress, but we need to do better." *The Lancet Infectious Diseases* 18(11): 1162-1163.

Nagraj, A. K. and D. Gokhale (2018). "Bacterial biofilm degradation using extracellular enzymes produced by *Penicillium janthinellum* EU2D-21 under submerged fermentation." *Advances in Microbiology* 8(9): 687-698.

Nahar, S., A. J.-w. Ha, K.-H. Byun, M. I. Hossain, M. F. R. Mizan and S.-D. Ha (2021). "Efficacy of flavourzyme against *Salmonella Typhimurium*, *Escherichia coli*, and *Pseudomonas aeruginosa* biofilms on food-contact surfaces." *International Journal of Food Microbiology* 336: 108897.

Negrut, N., S. A. Khan, S. Bungau, D. C. Zaha, C. Aron, O. Bratu, C. C. Diaconu and F. Ionita-Radu (2020). "Diagnostic challenges in gastrointestinal infections." *Rom J Mil Med* 123: 83-90.

Nicolle, L. E. (2014). "Catheter associated urinary tract infections." *Antimicrobial Resistance and Infection Control* 3(1): 23.

Nouha, K., R. S. Kumar, S. Balasubramanian and R. D. Tyagi (2018). "Critical review of EPS production, synthesis and composition for sludge flocculation." *Journal of Environmental Sciences* 66: 225-245.

Olsen, I. (2015). "Biofilm-specific antibiotic tolerance and resistance." *European Journal of Clinical Microbiology & Infectious Diseases* 34(5): 877-886.

Pagán, R. and D. García-Gonzalo (2015). "Influence of environmental factors on bacterial biofilm formation in the food industry: A review." *Journal of Postdoctoral Research* (ART-2015-95845).

Panebianco, F., F. Giarratana, A. Caridi, R. Sidari, A. De Bruno and A. J. L. Giuffrida (2021). "Lactic acid bacteria isolated from traditional

Italian dairy products: Activity against *Listeria monocytogenes* and modelling of microbial competition in soft cheese." *LWT* 137: 110446.

Passos da Silva, D., M. L. Matwichuk, D. O. Townsend, C. Reichhardt, D. Lamba, D. J. Wozniak and M. R. Parsek (2019). "The *Pseudomonas aeruginosa* lectin LecB binds to the exopolysaccharide Psl and stabilizes the biofilm matrix." *Nature Communications* 10(1): 2183.

Prado-Silva, L., V. Cadavez, U. Gonzales-Barron, A. C. B. Rezende, A. S. Sant'Ana (2015). "Meta-analysis of the effects of sanitizing treatments on *Salmonella, Escherichia coli O157: H7,* and *Listeria monocytogenes* inactivation in fresh produce." *Applied and Environmental Microbiology* 81(23): 8008-8021.

Qais, F. A., I. Ahmad, M. Altaf and S. H. Alotaibi (2021). "Biofabrication of Gold Nanoparticles Using *Capsicum annuum* Extract and Its Antiquorum Sensing and Antibiofilm Activity against Bacterial Pathogens." *ACS Omega* 6(25): 16670-16682.

Rabin, N., Y. Zheng, C. Opoku-Temeng, Y. Du, E. Bonsu, H. O. Sintim (2015). "Biofilm formation mechanisms and targets for developing antibiofilm agents." *Future Medicinal Chemistry* 7(4): 493-512.

Rajivgandhi, G. N., C. C. Kanisha, S. Vijayakumar, N. S. Alharbi, S. Kadaikunnan, J. M. Khaled, K. F. Alanzi and W.-J. Li (2021). "Enhanced anti-biofilm activity of facile synthesized silver oxide nanoparticles against K. *pneumoniae*." *Journal of Inorganic and Organometallic Polymers and Materials*: 1-13.

Rajivgandhi, G. N., G. Ramachandran, M. Maruthupandy, N. Manoharan, N. S. Alharbi, S. Kadaikunnan, J. M. Khaled, T. N. Almanaa and W.-J. Li (2020). "Anti-oxidant, anti-bacterial and anti-biofilm activity of biosynthesized silver nanoparticles using *Gracilaria corticata* against biofilm producing K. pneumoniae." *Colloids and Surfaces A: Physicochemical and Engineering Aspects* 600: 124830.

Rajwar, A., P. Srivastava and M. Sahgal (2016). "Microbiology of Fresh Produce: Route of Contamination, Detection Methods, and Remedy." *Critical Reviews in Food Science and Nutrition* 56(14): 2383-2390.

Ravindran, D., S. Ramanathan, K. Arunachalam, G. Jeyaraj, K. Shunmugiah and V. Arumugam (2018). "Phytosynthesized silver nanoparticles as antiquorum sensing and antibiofilm agent against the nosocomial pathogen *Serratia marcescens*: an *in vitro* study." *Journal of Applied Microbiology* 124(6): 1425-1440.

Richter, A. M., T. L. Povolotsky, L. H. Wieler and R. Hengge (2014). "Cyclic-di-GMP signalling and biofilm-related properties of the Shiga toxin-producing 2011 German outbreak *Escherichia coli* O104: H4." *EMBO Molecular Medicine* 6(12): 1622-1637.

Rosa, J. V. d., N. V. d. Conceição, R. d. C. d. S. d. Conceição and C. D. Timm (2018). "Biofilm formation by Vibrio parahaemolyticus on different surfaces and its resistance to sodium hypochlorite." *Ciência Rural* 48.

Roy, P. K., M. F. R. Mizan, M. I. Hossain, N. Han, S. Nahar, M. Ashrafudoulla, S. H. Toushik, W.-B. Shim, Y.-M. Kim and S.-D. Ha (2021). "Elimination of *Vibrio parahaemolyticus* biofilms on crab and shrimp surfaces using ultraviolet C irradiation coupled with sodium hypochlorite and slightly acidic electrolyzed water." *Food Control* 128: 108179.

Rubini, D., S. F. Banu, P. Nisha, R. Murugan, S. Thamotharan, M. J. Percino, P. Subramani and P. Nithyanand (2018). "Essential oils from unexplored aromatic plants quench biofilm formation and virulence of Methicillin resistant *Staphylococcus aureus*." *Microbial Pathogenesis* 122: 162-173.

Ryu, J.-H. and L. R. Beuchat (2005). "Biofilm formation by *Escherichia coli* O157:H7 on stainless steel: effect of exopolysaccharide and curli production on its resistance to chlorine." *Applied and Environmental Microbiology* 71(1): 247-254.

Saxena, P., Y. Joshi, K. Rawat, and R. Bisht (2019). "Biofilms: architecture, resistance, quorum sensing and control mechanisms." *Indian Journal of Microbiology* 59(1): 3-12.

Schulze, A., F. Mitterer, J. P. Pombo and S. Schild (2021). "Biofilms by bacterial human pathogens: Clinical relevance - development, composition and regulation-therapeutical strategies." *Microbial cell (Graz, Austria)* 8(2): 28-56.

Schulze, A., F. Mitterer, J. P. Pombo and S. J. M. C. Schild (2021). "Biofilms by bacterial human pathogens: Clinical relevance-development, composition and regulation-therapeutical strategies." *Microbial Cell* 8(2): 28.

Seper, A., V. H. Fengler, S. Roier, H. Wolinski, S. D. Kohlwein, A. L. Bishop, A. Camilli, J. Reidl and S. Schild (2011). "Extracellular nucleases and extracellular DNA play important roles in *Vibrio cholerae* biofilm formation." *Molecular Microbiology* 82(4): 1015-1037.

Shi, X. and X. Zhu (2009). "Biofilm formation and food safety in food industries." *Trends in Food Science & Technology* 20(9): 407-413.

Short, B., S. Carson, A.-C. Devlin, J. A. Reihill, A. Crilly, W. MacKay, G. Ramage, C. Williams, F. T. Lundy and L. P. J. C. R. i. M. McGarvey (2021). "Non-typeable *Haemophilus influenzae* chronic colonization in chronic obstructive pulmonary disease (COPD)." *Critical Reviews in Microbiology* 47(2): 192-205.

Sivaranjani, M., R. Srinivasan, C. Aravindraja, S. Karutha Pandian and A. Veera Ravi (2018). "Inhibitory effect of α-mangostin on *Acinetobacter baumannii* biofilms–an *in vitro* study." *Biofouling* 34(5): 579-593.

Soto González, S. M. (2014)."Importance of biofilms in urinary tract infections: new therapeutic approaches." *Advances in Biology* 1.

Soto, S., A. Smithson, J. Horcajada, J. Martinez, J. Mensa, J. J. C. M. Vila and infection (2006). "Implication of biofilm formation in the persistence of urinary tract infection caused by uropathogenic *Escherichia coli*." *Clinical Microbiology and Infection* 12(10): 1034-1036.

Steenackers, H., K. Hermans, J. Vanderleyden and S. C. De Keersmaecker (2012). "*Salmonella* biofilms: an overview on occurrence, structure, regulation and eradication." *Food Research International* 45(2): 502-531.

Tan, L., H. Li, B. Chen, J. Huang, Y. Li, H. Zheng, H. Liu, Y. Zhao and J. J. Wang (2021). "Dual-species biofilms formation of *Vibrio parahaemolyticus* and *Shewanella putrefaciens* and their tolerance to photodynamic inactivation." *Food Control* 125: 107983.

Tao, J., S. Yan, C. Zhou, Q. Liu, H. Zhu and Z. Wen (2021). "Total flavonoids from *Potentilla kleiniana* Wight et Arn inhibits biofilm formation and virulence factors production in methicillin-resistant *Staphylococcus aureus* (MRSA)." *Journal of Ethnopharmacology 278*: 114383.

Ternhag, A., A. Törner, A. Svensson, K. Ekdahl and J. Giesecke (2008). "Short- and long-term effects of bacterial gastrointestinal infections." *Emerging Infectious Diseases* 14(1): 143-148.

Uddin, M. S., M. Rahman, M. Faruk, A. Talukder, M. Hoq, S. Das and K. M. Islam (2021). "Bacterial gastroenteritis in children below five years of age: a cross-sectional study focused on etiology and drug resistance of *Escherichia coli* O157:H7, *Salmonella spp*., and *Shigella spp*." *Bulletin of the National Research Centre* 45(1): 1-7.

Vanaraj, S., B. B. Keerthana and K. Preethi (2017). "Biosynthesis, characterization of silver nanoparticles using quercetin from *Clitoria ternatea* L to enhance toxicity against bacterial biofilm." *Journal of Inorganic and Organometallic Polymers and Materials* 27(5): 1412-1422.

Vazquez-Armenta, F., A. Bernal-Mercado, M. Tapia-Rodriguez, G. Gonzalez-Aguilar, A. Lopez-Zavala, M. Martinez-Tellez, M. Hernandez-Oñate and J. Ayala-Zavala (2018). "Quercetin reduces adhesion and inhibits biofilm development by *Listeria monocytogenes* by reducing the amount of extracellular proteins." *Food Control* 90: 266-273.

Vázquez-Sánchez, D., J. Antunes Galvão and M. Oetterer (2018). "Contamination sources, biofilm-forming ability and biocide resistance of Shiga toxin-producing *Escherichia coli* O157:H7 and non-O157 isolated from tilapia-processing facilities." *Journal of Food Safety* 38(3): e12446.

Velázquez-Ordoñez, V., B. Valladares-Carranza, E. Tenorio-Borroto, M. Talavera-Rojas, J. A. Varela-Guerrero, J. Acosta-Dibarrat, F. Puigvert, L. Grille, Á. G. Revello and L. Pareja (2019). "Microbial contamination in milk quality and health risk of the consumers of raw milk and dairy products." *Nutrition in Health and Disease-our Challenges Now and Forthcoming time.*:181-217

Vijayakumar, K. and R. Thirunanasambandham (2021). "5-Hydroxy-methylfurfural inhibits *acinetobacter baumannii* biofilms: an *in vitro* study." *Archives of Microbiology* 203(2): 673-682.

von Rosenvinge, E. C., G. A. O'May, S. Macfarlane, G. T. Macfarlane and M. E. Shirtliff (2013). "Microbial biofilms and gastrointestinal diseases." *Pathogens and Disease* 67(1): 25-38.

Wang, H.-H., K.-P. Ye, Q.-Q. Zhang, Y. Dong, X.-L. Xu and G.-H. Zhou (2013). "Biofilm formation of meat-borne *Salmonella enterica* and inhibition by the cell-free supernatant from *Pseudomonas aeruginosa*." *Food Control* 32(2): 650-658.

Weerasooriya, G., A. R. McWhorter, S. Khan and K. K. Chousalkar (2021). "Transcriptomic response of *Campylobacter jejuni* following exposure to acidified sodium chlorite." *Science of Food* 5(1): 1-9.

WHO. (2020, April 20, 2021). *Food Safety*. Retrieved September 7, 2021, from https://www.who.int/news-room/fact-sheets/detail/food-safety.

WHO, W. H. O. (2017). *Diarrhoeal disease. Fact Sheet* Retrieved July 24, 2021, from https://www.who.int/news-room/fact-sheets/detail/diarrhoeal-disease.

Wickramasinghe, N. N., M. M. Hlaing, J. T. Ravensdale, R. Coorey, P. S. Chandry and G. A. Dykes (2020). "Characterization of the biofilm matrix composition of psychrotrophic, meat spoilage pseudomonads." *Scientific Reports* 10(1): 16457.

Yan, L., W. Wu and S. Tian (2020). "Antibacterial and antibiofilm activities of *Trollius altaicus* C. A. Mey. On *Streptococcus mutans*." *Microbial Pathogenesis* 149: 104265.

Yan, X., S. Gu, Y. Shi, X. Cui, S. Wen and J. Ge (2017). "The effect of emodin on *Staphylococcus aureus* strains in planktonic form and biofilm formation *in vitro*." *Archives of Microbiology* 199(9): 1267-1275.

Yang, X., G. N. Rajivgandhi, G. Ramachandran, N. S. Alharbi, S. Kadaikunnan, J. M. Khaled, T. N. Almanaa and N. Manoharan (2020). "Preparative HPLC fraction of *Hibiscus rosa*-sinensis essential oil against biofilm forming *Klebsiella pneumoniae*." *Saudi Journal of Biological Sciences* 27(10): 2853-2862.

Yang, Y., X. Yu, L. Zhan, J. Chen, Y. Zhang, J. Zhang, H. Chen, Z. Zhang, Y. Zhang and Y. Lu (2017). "Multilocus sequence type profiles of *Bacillus cereus* isolates from infant formula in China." *Food Microbiology* 62: 46-50.

Yu, H., X. He, W. Xie, J. Xiong, H. Sheng, S. Guo, C. Huang, D. Zhang and K. Zhang (2014). "Elastase LasB of *Pseudomonas aeruginosa* promotes biofilm formation partly through rhamnolipid-mediated regulation." *Canadian Journal of Microbiology* 60(4): 227-235.

Zhang, C., C. Li, M. A. Abdel-Samie, H. Cui and L. Lin (2020). "Unraveling the inhibitory mechanism of clove essential oil against *Listeria monocytogenes* biofilm and applying it to vegetable surfaces." *LWT* 134: 110210.

Zhang, J. and C. L. Poh (2018). "Regulating exopolysaccharide gene wcaF allows control of *Escherichia coli* biofilm formation." *Scientific Reports* 8(1): 13127.

Zhang, N., L. Wang, X. Deng, R. Liang, M. Su, C. He, L. Hu, Y. Su, J. Ren, F. Yu, L. Du and S. Jiang (2020). "Recent advances in the detection of respiratory virus infection in humans." *Journal of Medical Virology* 92(4): 408-417.

Zhao, X., F. Zhao, J. Wang and N. Zhong (2017). "Biofilm formation and control strategies of foodborne pathogens: food safety perspectives." *RSC advances* 7(58): 36670-36683.

Zogaj, X., M. Nimtz, M. Rohde, W. Bokranz and U. Römling (2001). "The multicellular morphotypes of *Salmonella typhimurium* and *Escherichia coli* produce cellulose as the second component of the extracellular matrix." *Molecular Microbiology* 39(6): 1452-1463.

In: Pathogenic Bacteria
Editor: Keith D. Watts
ISBN: 978-1-68507-422-7

Chapter 2

BIOPHYSICAL TOOLS TO EXPLORE THE ANTI-VIRULENCE MODE OF ACTION OF PHYTOCHEMICALS AGAINST PATHOGENIC BACTERIA

F. J. Vázquez-Armenta[1], A. A. López-Zavala[1], A. T. Bernal-Mercado[2], M. R. Tapia-Rodríguez[3], D. Encinas-Basurto[4] and J. F. Ayala-Zavala[5,*]

[1]Departamento de Ciencias Químico Biológicas, Universidad de Sonora, Hermosillo, Sonora, México

[2]Departamento de Investigación y Posgrado en Alimentos, Universidad de Sonora, Hermosillo, Sonora, Mexico

[3]Departamento de Biotecnología y Ciencias Alimentarias, Instituto Tecnológico de Sonora, Ciudad Obregón, Sonora, México.

[*] Corresponding Author's E-mail: jayala@ciad.mx.

[4]Departamento de Agricultura y Ganadería,
Universidad de Sonora, Hermosillo, Sonora, México
[5]Coordinacion de Tecnología de Alimentos de Origen Vegetal,
Centro de Investigación en Alimentación y Desarrollo,
Colonia la Victoria, Hermosillo, Sonora, Mexico

Abstract

Antibiotic resistance is one of the major threats to global health; therefore, the challenge is developing strategies to attack this problem. Anti-virulence therapy is an alternative that has gained attention in the last years, and it is aimed to attenuate the bacterial strategies to infect and cause disease, and sometimes it does not aim to affect the pathogen viability. Phytochemicals have been demonstrated anti-virulence properties against a wide range of pathogenic bacteria. Studies on this topic are commonly focused on the characterization of physiological changes of treated bacteria such as bacterial membrane damage, biofilm inhibition or changes in expression of virulence-related genes. However, few studies delve into the molecular targets of these compounds and their molecular interactions. In this sense, biophysical techniques that include spectroscopy methods (X-ray crystallography, UV-Vis, IR, fluorescence, and others), surface plasmon resonance (SPR), isothermal titration calorimetry (ITC), and nuclear magnetic resonance (NMR) can help to validate the action sites of molecular targets of natural compounds. These tools have been used for decades in drug design in the pharmaceutical industry since they allow a detailed mechanistic characterization of compound binding. This information is quite useful when studying the mechanisms of action of natural plant compounds. Therefore, this chapter provides an overview of biophysical techniques that can validate targets of phytochemicals with anti-virulence properties that can help elucidate their mode of action.

Keywords: natural compounds, virulence factors, natural antimicrobials, molecular interactions.

INTRODUCTION

The spread of antibiotic-resistant bacteria has increased in the past years, driving the search for novel compounds with antimicrobial properties (Nakano et al. 2015, Grundner et al. 2007). Therefore, treatment of infections caused by antibiotic-resistant or multi-antibiotic resistant bacteria requires the use of very specific antibiotics, but its development is not economically viable for pharmaceutical companies (Simpkin et al. 2017). Thus, the implementation of novel strategies to combat the spread of antibiotic resistance has gained significant importance.

One of the most promising strategies is the anti-virulence therapy aimed to interfere with pathogenic bacteria's infectious processes, attenuating the production of virulence factors (Storz et al. 2013). Virulence factors are any bacterial product or strategy involved in establishing infection (Sui et al. 2009). This definition includes bacterial toxins, cell surface proteins that mediate the bacterial attachment, cell surface carbohydrates and proteins that protect a bacterium, and hydrolytic enzymes that may contribute to the pathogenicity of the bacterium and biofilm formation to survive within the host (Camejo et al. 2011, Peterson 1996). Another characteristic of anti-virulence therapy is that it is achieved without affecting the bacterial viability, reducing the selective pressure exerted by conventional antibiotics focused on interrupting the vital process of bacteria or inhibiting its replication.

Plants are an essential source of diverse compounds that possess antimicrobial and anti-virulence properties against a wide range of pathogenic bacteria. Traditional medicine uses plant extracts or infusions to combat various infectious diseases, and these properties have been attributed to the presence of phytochemicals (Choudhury et al. 2020). These compounds are part of the secondary metabolism of plants synthesized in response to environmental cues, and many of them are related to plant defense against pathogens or tolerance to stressful conditions (Franzoni et al. 2019). Phytochemicals are classified according to their chemical and structural characteristics in phenolic compounds (phenolic acids, flavonoids, tannins, and stilbenes), terpenes, sulfur compounds, alkaloids, saponins, lignins, among others (Suresh and Abraham 2020).

The antimicrobial properties of phytochemicals have been demonstrated extensively in the scientific literature. These properties are presented in a concentration-dependent manner and are related to bacterial replication inhibition or bacterial cell inactivation (Omojate Godstime et al. 2014). However, at lower concentrations where antimicrobial properties are not exerted, the anti-virulence properties are observed. Examples of this therapy could be the inhibition of cell-to-cell communication necessary for accurate regulation of virulence gene expression, impairment of biofilm development, inhibition of bacterial toxins production, inhibition of adhesion and invasion of host cells, or impairment of regular infection cycles (Bernal-Mercado et al. 2018, Tapia-Rodriguez et al. 2019, Vazquez-Armenta et al. 2020). Together with their harmless for human health and the lower or absent selective pressure, these properties make them candidates for their use in anti-virulence therapy.

Despite the evidence of the phytochemicals' effectiveness for attenuating pathogenic bacteria virulence, there are no treatments based on their use. For the approval of natural compounds to be used as therapeutic agents, candidates are required to pass multiple safety tests and clinical trials (Genick and Wright 2017, Matthews et al. 2016). One of the main impediments of phytochemicals to move to the subsequent phases of development is the lack of knowledge of their specific site of action (Dai et al. 2020). Studies describing the effect of these compounds on the morphology and physiology of treated bacteria, but in some cases, the molecular target is not validated. Thus, to expand the availability of natural anti-virulence agents, the elucidation of their mode of action is required, starting with the validation of their molecular targets (Beckham and Roe 2014).

Biophysics is where physical theories explain biological processes, including the interaction between molecules (Campbell 2012). In this sense, biophysical techniques can be helpful in the identification and validation of molecular targets and understanding molecular interactions. Biophysical tools such as X-ray crystallography, spectroscopy (UV-Vis, infrared, and fluorescence), surface plasmon resonance (SPR), isothermal titration calorimetry (ITC), nuclear magnetic resonance (NMR), among others, have become critical components of drug discovery and development in the pharmaceutical industry (Renaud et al. 2016). These techniques allow obtaining specific information about the interaction of small molecules with large biomolecules, binding energy, binding affinity parameters, and conformational changes in target biomolecules (Folmer 2016). This information helps establish the mechanisms of action and supports the guided design of more efficient compounds (Genick and Wright 2017). Therefore, this chapter discusses the basis of biophysical tools that can explain the mode of action of phytochemicals with anti-virulence properties.

Biophysical Methods to Evidence the Interaction of Phytochemicals with Lipid Membrane

Langmuir Monolayers

Biological membranes are highly complex, which limits their experimental studies (Elderdfi and Sikorski 2018). Therefore, the Langmuir monolayer technique is an excellent alternative to obtain high-quality ordered monolayers that mimic cell membranes and thus explore the characteristics of membrane structure and their interaction with phytochemicals (Nowotarska et al. 2014, Stefaniu et al. 2014). This technique has been successfully used to study protein and drugs interaction with lipid models (Hac-Wydro and Dynarowicz-Latka 2008, Boisselier et al. 2017). The method consists of the transfer process from the air-water interface onto a solid substrate of a monomolecular layer adsorbed at the air-water interface (Velázquez et al. 2016). This approach is based on the spread of amphiphilic molecules distributed on a water surface to form an insoluble monolayer at an interface (de Carvalho et al. 2019). For this, the amphiphilic material is dissolved in a water-insoluble volatile solvent, dropped on the air-water interface, and the monolayer is formed when the solvent evaporates (Hac-Wydro and Dynarowicz-Latka 2008). The molecules used to build monolayers have a polar end immerse in water and a long hydrophobic chain oriented towards the air. These amphiphilic molecules can be composed of the phospholipids present in bacterial membranes and must have a long hydrocarbon chain to form the insoluble monolayer.

The deposited lipid solution will be distributed to cover the entire available area. If the site is large, the distance between adjacent molecules will be large and their interactions weak. When the monolayer reaches the thermodynamic equilibrium, it can be compressed using two barriers (Langmuir-Blodgett balance, Figure 1) to reduce the surface area (Velázquez et al. 2016). During the compression, the surface pressure and the molecular area are constantly monitored to obtain an isotherm. Surface pressure is measured by the Wilhelmy plate method, which measures the force due to surface tension on the suspended plate partially immersed in the subphase.

The isothermal compression changes the structure of the monomolecular film, which passes through a series of two-dimensional states, referred to as gas, expanded and compressed liquids, and solid states (Figure 2) (Petty 1996). When the molecules are in the subphase, the monolayer exists in the gaseous state, and with compression, it can reach a transition phase to a liquid-expanded state. After compression, it transitions to a liquid-condensate state, and the monolayer reaches the solid-state (Petty 1996). If the monolayer is compressed after reaching the solid-state, the monolayer will collapse into three-dimensional structures. The understanding of the 2D phase diagram of the monolayer helps to associate its physical and chemical properties (Velázquez et al. 2016). In addition, Langmuir's technique can be used to prepare highly organized multilayers by the successive immersion of a solid substrate called Langmuir-Blodgett.

The Langmuir monolayer technique can provide helpful information regarding the interaction of phytochemicals with model membranes and the effect on physical characteristics like surface pressure, isotherm behavior, and molecule packing density (Ferreira et al. 2016). Changes in Langmuir isotherms may explain the influence of bioactive compounds on the lipid molecules arrangement compared to control isotherms.

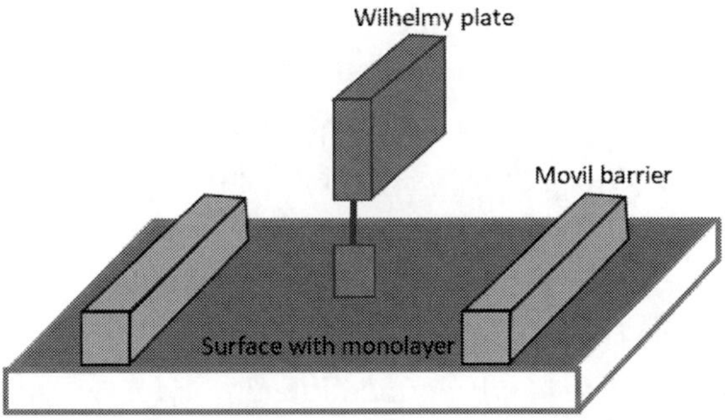

Figure 1. Model of Langmuir-Blogett balance.

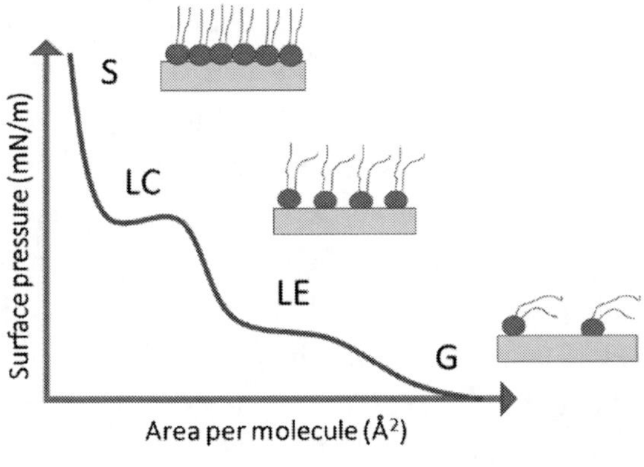

Figure 2. Surface pressure – area (π-A) isotherms of monolayer and orientation of molecules in different phases. G = gaseous state, LE = liquid-expanded state, LC = liquid-condensate state, S = solid state.

For example, the increase in lift-off values by the bioactive compounds in the subphases indicates that the monolayers occupied a larger area at low surface pressure suggesting the least efficient phospholipids molecular packing and, therefore, a more fluid membrane (Nowotarska et al. 2014). Moreover, a maximum value reduction of the compressibility modulus of the isotherms indicates an increase in the elasticity/fluidity of the membrane. On the contrary, higher values of the modulus of compressibility than the reference indicate a more significant interaction with lipids and, therefore, cause the membrane's rigidity. In addition, Langmuir balance can measure amphiphilic molecule's surface potential, which gives information concerning the orientation of films components (Ferreira et al. 2016).

Some studies have used the Langmuir balance to elucidate the effect of phytochemicals in bacterial membrane models and its relationship with anti-virulence properties. For example, Bernal-Mercado et al. (2020) showed that vanillic acid, protocatechuic acid, catechin, and their combination inhibited the adhesion of uropathogenic *Escherichia coli* (UPEC) to silicone surface by impairing bacterial motility, reducing the production of fimbriae and extracellular polymeric substances, and modifying the bacterial membrane properties. To delve into the mechanism of action, the authors studied the interaction of phenolic compounds with Langmuir monolayers. They reported that vanillic acid, protocatechuic acid, and catechin disturbed the packing density of the phospholipid monolayer showing a fluidizing effect explained by the interactions between dissociated and negative charges of hydroxyl groups with the negative charge of 1,2-dimyristoyl-sn-glycerol-3-phosphoglycerol (DMPG) and 1,2-dipalmitoyl-sn-glycerol-phosphatidylcholine (DPPC), causing the monolayer expansion (Bernal-Mercado et al. 2020). Similarly, Ferreira et al. (2016) found that thymol, a terpenoid, could expand DPPC monolayers, decrease their surface elasticity, and change the morphology of the lipid monolayer by its incorporation lipid Langmuir membrane.

In the same field, Nowotarska et al. (2014) used the Langmuir technique to study the interaction of carvacrol, cinnamaldehyde, geraniol, 2,5-di-hydroxybenzaldehyde and 2-hydroxy-5-methoxy-benzaldehyde with membrane models. These monolayers were composed of phospholipids of a zwitterionic and anionic nature that predominate in bacterial cells like 1,2-di-hexadecanoyl-sn-glycero-3-phosphoethanolamine (DPPE), 1,2-dihexadecanoyl-sn-glycero-3-phospho-(1'-rac-glycerol) (DPPG), and 1,1',2,2'-tetra-tetra-decanoylcardiolipin (cardiolipin). The results showed that depending on their structure, phytochemicals could modify the cell membrane by incorporating them into the lipid monolayer, reducing the effectiveness of molecular packaging, increasing their fluidity, forming aggregates, and modifying the dipole moment of the monolayers. Furthermore, the authors found an influence on the nature of the lipid monolayer. They found that dissociated and negative hydroxyl groups can interact with the positive charges of the amino group of lipids through electrostatic forces making the lipid molecules closer together increasing lipid packing. While to increase elasticity/fluidity, phytochemicals were incorporated into the hydrophobic region of lipids by lipophilic attraction and through electrostatic repulsion between the compound and the negatively charged polar heads. The authors suggested that the hydrophilic and ionic part associated with the OH groups of phenol and the lipophilic part of the benzene group facilitate interactions with the monolayer.

We have only reviewed the fundamental Langmuir balance experiments, but several complementary studies could be developed to extend the monolayer technique possibilities, such as spectroscopic methods, fluorescence microscopy, Brewster angle microscopy, infrared spectroscopy, X-ray diffraction, or atomic force microscopy (Maget-Dana 1999). The Langmuir monolayer technique provides a successful model membrane to study its physicochemical characteristics and an insight into the interaction of anti-virulence compounds on bacterial cell membranes. However, literature regarding the interaction of phytochemicals with lipid Langmuir monolayers is scarce, and more studies could be conducted on this approach.

Fluorescence Polarization

Fluorescence polarization is a valuable tool to provide biophysical information on molecular orientation and mobility, including membrane lipid fluidity using a hydrophobic fluorescent dye that intercalates into the fatty acid bilayer (Figure 3) (Tiwari et al. 2021). This approach is gaining attention for studying the fluidity of bacterial membranes and for drug discovery because of its sensitivity, specificity, availability of probes, and the simplicity of detection (Trevors 2003). In addition, fluorescent probes extend several advantages; they are helpful to be used with living cells without a pre-treatment, can measure static and dynamic components of membrane fluidity, and can perform a virtual real-time measure of cytoplasmic membrane fluidity (Mykytczuk et al. 2007).

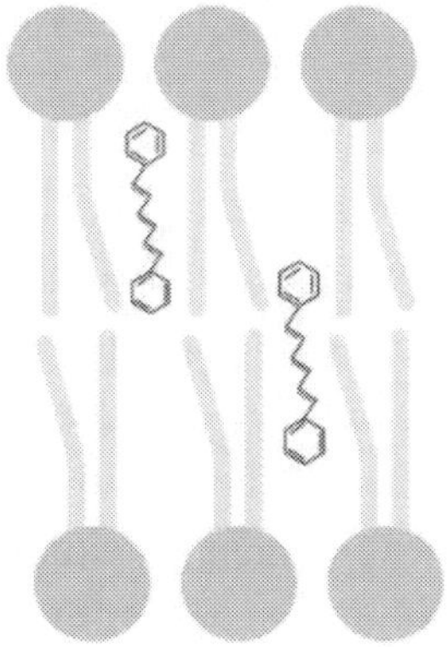

Figure 3. DPH fluorescent probe in the membrane bilayer.

This technique consists of a fluorophore excited using a spectrofluorometer with a UV light polarized as a vertical excitation beam and passed across a polarizing filter and the dye-labeled membrane (Lea and Simeonov 2011). The degree of polarization is measured through the emission of parallel ($F\|$) and horizontal ($F\perp$) beams perpendicular to the excitation light and is expressed in terms of fluorescence polarization or anisotropy as follows:

$$P = (F\|)/(F\perp)$$

The fluorescent probes are either totally or partially lipophilic, and they are incorporated into a specific membrane region with little membrane disruption (Trevors 2003). Different probes have been employed, but the most often used fluorescent probe for labeling membranes is diphenylhexatriene (DPH). DPH is a hydrophobic symmetrical, and rod-like trans-polyene that inserts in the membrane's hydrophobic fatty acid chains and rotates in the lipid lattice at different rates depending on the membrane's compactness (Tiwari et al. 2021).

Fluorescence polarization is widely used to study the physiological state of bacteria under environmental conditions such as the presence of phytochemicals (Trevors 2003). Membrane fluorescent probes are used in this procedure, and their emission in different directions fluctuates depending on their tumbling movements inside the lipid bilayer (Selvaraj et al. 2015). Any molecule that limits the movements of the lipid chains increases the rigidity of the membrane, observing a reduction in anisotropy emission. In contrast, those that interact with the membrane and increase its fluidity will be characterized by increased fluorescence (Ionescu et al. 2013).

This approach can be used for screening phytochemicals with an effect on the bacterial membrane. For example, the interaction of flavonoids with membrane models was evaluated by fluorescence polarization (Wu et al. 2013). The authors used large unilamellar vesicles formed by DPPC and DPPG to simulate Gram-negative bacteria membranes (Wu et al. 2013). Flavonoids such as kaempferol, chrysin, quercetin, baicalein, and luteolin increased polarization values resulting in membrane rigidification, while lower polarization values were found for isoflavonoids like puerarin, ononin, daidzein, genistin, and tangeritin, increasing membrane fluidity. Results showed that this interaction is influenced by the number and position of hydroxyl groups. The hydroxyl group at C-3 in the ring C in flavonoids is crucial for decreasing membrane fluidity. These compounds can be incorporated into the interior of lipid bilayers and increase the ordering and dynamics in the membrane interior. At the same time, isoflavonoids that do not contain hydrophobic prenyl chains interact with the polar region of lipid bilayer or penetrate the polar/apolar interface with the hydrophobic core not directly affected (Wu et al. 2013).

Similarly, fluorescence polarization measurements demonstrated that flavonoids (1–10 M) interacted on the deeper areas of lipid bilayers to decrease membrane fluidity in a structure-dependently manner finding that flavonols are more suitable to rigidify membranes than flavones (Tsuchiya 2010). This technique was employed to assess the membrane interaction among the flavanones hesperidin (glycoside) and hesperetin (aglycone), and dimyristoyl-phosphatidyl choline (DMPC) liposomes using fluorescence polarization. The results suggested that hesperidin is located near the polar head groups region, and the hesperetin interacts strongly with membrane acyl chains (Londoño-Londoño et al. 2010).

Another report employing the fluorescence anisotropy technique suggested that sophoraflavanone G isolated from *Sophora exigua* at 0.05-5 μg/mL corresponding to the effective concentration against several bacteria growth, increased fluorescence polarization of the liposomes formed by 1,2-dipalmitoyl-L-α-phosphatidyl-choline and l-palmitoyl-2-oleoyl-L-α-phosphatidylcholine. This compound diminished the fluidity of the outer and inner layers of membranes due to their ability to interact with the phospholipid's hydrophilic head and hydrophobic region (Tsuchiya and Iinuma 2000). However, some reports have demonstrated the opposite effect. Ionescu et al. (2013) evaluated the membrane interaction of quercetin on 2-dimyristoyl-sn-glycero-3-phosphocholine small unilamellar vesicles (SUVs) with different amounts of cholesterol, using Laurdan as a fluorescent probe. The findings revealed that quercetin increased the membrane fluidity in a dose-dependent manner, attributed to the hydrophobic/hydrophilic character of the substance.

The studies in this field have demonstrated the advantage of the fluorescence polarization method to understand the phytochemicals interaction in membrane models proving specific insights about their effect on membrane fluidity. The interaction of phytochemicals with the bacterial membrane is a crucial point to understand the anti-virulence mechanisms.

Biophysical Methods to Validate the Interaction of Phytochemicals with Virulence Proteins

It is necessary to identify and validate the phytochemical's site of action to propose them as anti-virulence agents (Dai et al. 2020). Among molecular targets that have been identified, proteins related to infectious processes are included. For example, proteins related to bacterial communication (synthases and receptors) such as *quorum-sensing*, transcriptional factors, adhesins involved in the adhesion of bacteria to host-cell, bacterial toxins that bind to the host cell membrane or enzymes that covalently anchor diverse virulence factors to the bacterial cell wall, among others (Díaz-Nuñez et al. 2021, Luna-Solorza et al. 2020). However, to understand the mode of action of natural anti-virulence compounds, it is necessary to obtain precise information about the binding and the induced conformational changes. In this sense, several biophysical tools can help to get this kind of parameter.

Tools Aimed to Determine Binding Affinity

X-Ray Crystallography

X-ray crystallography is a non-destructive technique used to analyze a wide range of solid-state materials, including minerals, metals, polymers, ceramics, and protein crystals (Ameh 2019). In the field of drug discovery and development in the pharmaceutical industry, this technique has evolved as a method of choice for accurate determination of molecular structure at atomic resolution as it has proven to be an invaluable tool to provide comprehensive structural information about the interaction of small molecules with target proteins (Figure 4) (Maveyraud and Mourey 2020, Aitipamula and Vangala 2017). X-ray crystallography can determine the structure of small molecules at the atomic level or identify the secondary structures, shapes, and motifs of proteins. The identification of secondary structures leads to determining protein's overall three-dimensional structure (Ooi 2010). For these reasons, X-ray crystallography has become a powerful tool for screening drug candidates, including natural compounds, as it can also provide structural information of protein-ligand complexes with diverse binding affinities and insights into the physical chemistry of the complex formation (Maveyraud and Mourey 2020).

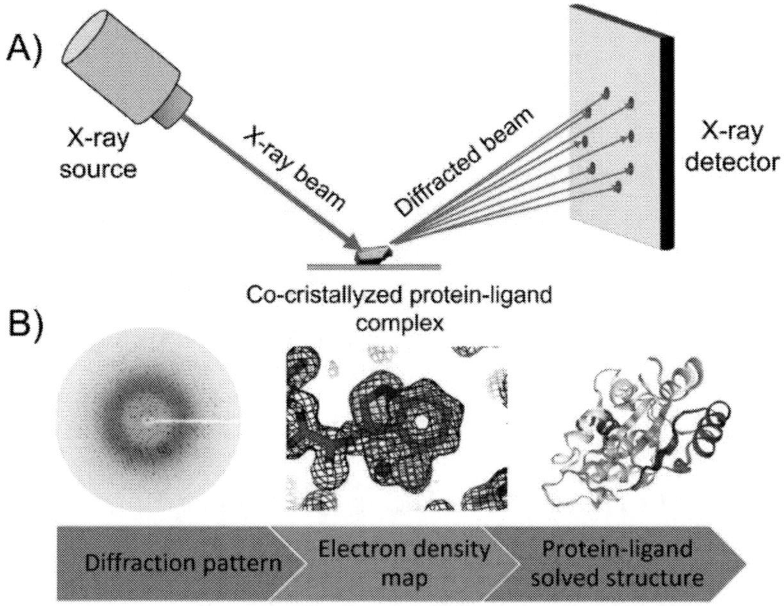

Figure 4. Schematic representation of X-ray crystallography experiment. A) Data acquisition (diffraction pattern) from the co-crystallized protein-ligand complex. B) Workflow for structure solving from the diffraction pattern.

The development of X-ray crystallography as a precise tool for the analysis of protein-ligand complexes is the result of pioneering works since the discovery of X-rays by W. C. Röentgen in 1895, the demonstration of the ability of X-rays to be diffracted by single crystals by M. von Laue in 1912 and the formulation of diffraction law by W. L. Bragg in 1915 (Aitipamula and Vangala 2017). In crystalline structures, the atoms are arranged in a regular and repeated pattern, and the smallest repeating element in the crystal is called the unit cell (Ameh 2019). When x-ray impinges upon atoms of crystallized protein, the x-rays are scattered by the electrons in the atoms. Thus, from the diffraction pattern, the position of each atom in the unit cell can be determined (Ooi 2010).

The target protein must be co-crystallized with the potential inhibitor to study their interactions. In this process, a screening of crystallization conditions should be considered as the presence of the compound might alter the crystallization conditions of the protein alone (Maveyraud and Mourey 2020). Once the cocrystal is obtained, an X-ray crystallographic experiment can be carried out to obtain the electron density map from the diffraction pattern with standard data processing software. Then, the atoms must be fitted into the electron density map and refined. At this point, the correct interpretation of whether electron density observed in a binding site is compatible with ligand or water of buffer molecules is required (Pearce et al., 2017). This situation is challenging when the studied compounds are small molecules, obtaining low-resolution density maps (Smart et al. 2018). These challenges can be overcome with high-resolution data, visual inspection of the ligand/protein model, and electron-density maps and ligand validation tools (Smart et al. 2018).

Valuable information obtained from natural compounds co-crystallized with target proteins is the binding mode that helps elucidate the molecular mechanisms underlying the recognition of ligands and inhibitors. The correct binding mode cannot be determined with certainty using molecular simulations; thus, X-ray crystallography is the preferred tool for this purpose. The details of molecular interaction at the atomic level related to specific ligand recognition are helpful to understand the phenomenon's principles (Kasahara and Kinoshita 2016). The ligand's binding mode or molecular orientation helps identify functional groups important for inhibiting targeted protein activity. For example, Steinbach et al. (2008) obtained the X-ray structure of the anti-virulence target enzyme MurA, present in various pathogenic bacteria, complexed with its substrate and the sesquiterpene lactone cnicin from herbal origin. The structure-activity analysis and biochemical assays indicated that this compound acted as an irreversible inhibitor and identified the unsaturated side chain coupled to the macrocycle lactone that contains the carbonyl function that acts as a substrate-mimicking, electrophilic agent binding to Cys115 of MurA.

Additionally, the use of practical information on binding mode allows searching other natural compounds with a similar structure that can be more efficient inhibitors, or this information can be used to improve the binding affinity of the natural compound with chemical substitutions in the called structure-guided drug design (Jacquemard et al. 2019). This approach has been followed to design carbohydrate-based inhibitors of lectins LecA and LecB of *Pseudomonas aeruginosa* related to establishing life-threatening chronic infections through biofilm formation (Titz 2014).

Similarly, Sommer et al. (2015) synthesized and evaluated more than 20 derivates from previously identified glycomimetic cinnamides inhibitors of LecB. The structure-activity relationship allowed the experimental determination of the binding mode of these cinnamides with LecB. It revealed that cinnamide substituent forms lipophilic interactions with Glt97 and Thr98 and one water-mediated hydrogen bond via the carbonyl oxygen with Ser23 of LecB from *P. aeruginosa*. Kalas et al. (2018) used X-ray structure-guided methods and complementary assays to design aryl galactosides and N-acetylgalactosaminosides to inhibit the F9 pilus adhesin FmlH used by uropathogenic *E. coli* during urinary tract infection. The obtained compound named 29β-NAc presented a higher affinity than regular carbohydrate–lectin interactions (Kalas et al. 2018). These studies illustrate how X-ray crystallography experiments help to increase the repertory of anti-virulence compounds to fight bacterial infections.

Isothermal Calorimetry (ITC)

Isothermal titration calorimetry (ITC) is a label-free binding assay that measures heat transferred in molecular interactions. This biophysical technique is widely used to determine the binding affinity, stoichiometry, and thermodynamics of the interaction of various biological reactions and biotechnological applications (Roselin et al. 2010). In protein, science is used to validate protein-small molecules (ligands) interaction and lead the thermodynamic characterization in binding reactions (Liang 2008). It is considered the gold-standard technique for the complete thermodynamic profile determination giving a detailed understanding of the driving forces underlying the binding process (Menéndez 2021). ITC is the only technique that determines the enthalpy change occurring in molecular interaction directly from transferred heat. This information is critical in targeted drug design as it allows to obtain the intrinsic parameters of binding reactions (Baranauskiene et al. 2019). Additionally, ITC has been demonstrated to be a powerful tool to validate targets related to the virulence process in pathogenic bacteria (Kamal et al. 2018, Storz et al. 2013).

ITC instrument is based on the work of Keily and Hume (1956) that described how the heat of reaction might be estimated by the slope of thermometric titration and the pioneering work of Tyson et al. (1961), who described an apparatus for conducting thermometric titrations by measuring the difference in temperature between a reaction vessel and blank solution. Modern calorimeters contain the same fundamental pieces, a sample cell, a reference cell, and an injection syringe (Archer and Schulz 2020). In protein-ligand interaction experiments (Figure 5), the syringe containing a ligand is titrated into a cell containing the target protein. In a typical ITC experiment, the difference in temperature between the sample cell and the reference cell is monitored. The heat released or absorbed during binding reaction causes an imbalance between the reference cell and sample cell, and the equipment compensates for it by applying the power of the heaters (Chaires 2008). The energy transferred in the binding reaction is calculated from the power required to keep the temperature in the reaction cell equal to the temperature in the reference cell.

The released or absorbed heat in each injection is registered until the binding sites in the target protein become saturated to obtain the binding parameters. A titration curve is obtained by integrating the peaks of each binding event and plotting the integrations against the molar ratio (Figure 6). From the ITC titration curve, the change in enthalpy (ΔH), binding constant (Kd), and binding stoichiometry (n) can be obtained (Archer and Schulz 2020). Other thermodynamic parameters as the Gibbs energy of binding can be obtained from $\Delta G = -RT \ln Kd$, and the change in entropy (ΔS) can be determined from the relationship $\Delta G = \Delta H - T\Delta S$. These calculations are made using data analysis programs provided by the instrument manufacturer (Baranauskiene et al. 2019).

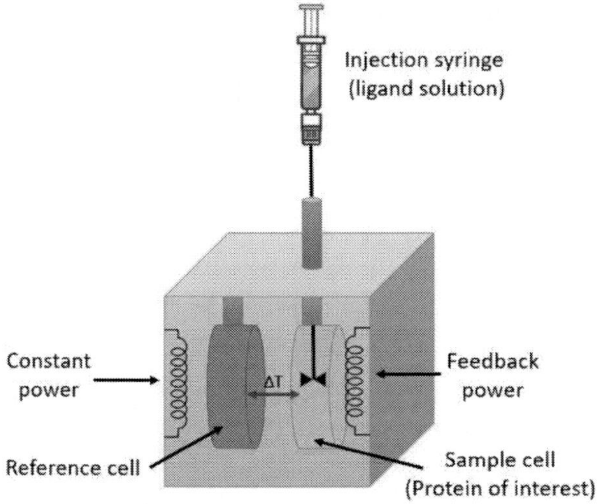

Figure 5. Schematic representation of an ITC instrument. Ligand is injected by the syringe into sample cell containing the protein of interest. The heat released or absorbed during binding reactions modifies the temperature of sample cell and a feedback circuit modifies the power applied to maintain the same temperature as the reference cell.

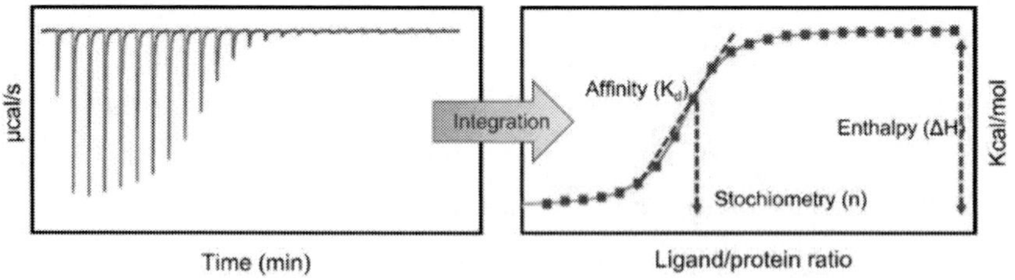

Figure 6. Typical thermogram (left) and binding isotherm (right) obtained from an ITC experiment. Each peak in the thermogram is the result of each injection of ligand solution into the sample cell. The intensity of each peak is proportional to the heat released (or absorbed) in binding reactions. Areas under each peak are calculated and normalized per mole of ligand and plotted against the molar ratio (right). From the resulted plot, the thermodynamic parameters of the binding reaction can be obtained.

ITC has been used to validate the binding of natural compounds with proteins related to virulence in pathogenic bacteria. Corral-Lugo et al. (2016) used ITC to validate the interaction of natural compounds with the *quorum-sensing* regulator RhlR of *P. aeruginosa* PAO1. The experimental approach consisted of identifying potential ligands by *in-silico* docking experiments and its validation with ITC measurements. As a result, the plant-derived phenolic compound rosmarinic acid was the only ligand able to bind to RhlR protein with a Kd = 0.49 mM, a favorable enthalpy changes (ΔH) of -0.4 kcal/mol and a similar stoichiometry to the natural ligand of the *quorum-sensing* regulator. The additional assays demonstrated that this compound acts as an agonist of the cell-to-cell communication system.

Li et al. (2016) investigated the inhibitory effect of the natural compound chalcone on Sortase A (SrtA) from *Listeria monocytogenes*, an important protein that anchors virulence proteins covalently to the cell wall in gram-positive bacteria. Using molecular simulations, site-directed mutagenesis, and ITC experiments, the authors characterized the binding mechanism of chalcone with SrtA and identified key residues indispensables for the inhibitory activity. ITC analysis showed that the interaction of chalcone with wild-type SrtA had a Kd = 0.112 mM, a ΔH = -0.86 kcal/mol and a ΔS = 15.2 cal/mol. The obtained results help to validate the results observed in the *in-vitro* and *in-vivo* infection models, demonstrating that targeting SrtA with natural compounds could be a useful strategy for preventing and treating *L. monocytogenes* infection.

Thermodynamic parameters describe if a binding reaction will occur spontaneously and provide insight into the mechanism of binding. The thermodynamic characterization of binding reactions is of enormous value in the study of natural compounds with anti-virulence properties as they provide information about the driving forces that cannot be obtained from other methods alone. This information helps understand natural compound's efficacy since it determines the equilibrium states that often may not be established *in-vivo* (Menéndez 2021, Chaires 2008). Decompose the free energy of binding in its enthalpic (ΔH) and entropic (ΔS) components provide insights into the molecular nature of the binding process (Menéndez 2021). This information has been used to optimize the binding energies of small organic compounds against virulence proteins (Zender et al. 2013, Storz et al. 2013, Nakano et al. 2015, Grundner et al. 2007). The use of ITC to validate the binding of natural compounds in target proteins is a practical approach since ITC experiments do not require immobilization or modification of proteins; modern instruments are simpler to use and require low maintenance. Nevertheless, special attention must be required in protein quality (purity, homogeneity, and integrity). Once the protein purification process has been optimized, automated methods can be implemented to search for potential candidates with anti-virulence potential.

Surface Plasmon Resonance (SPR)

Surface plasmon resonance (SPR) technology is a powerful tool for studying biomolecular interactions, such as protein-protein, protein-DNA, protein-carbohydrate, protein-lipid, and protein-small molecules interactions (Vachali et al. 2016). The SPR approach offers excellent benefits such as low sample consumption and reproducible and real-time information on ligand-binding interaction (Hodnik and Anderluh 2013). SPR is a label-free detection technology that has evolved as a suitable and dependable platform in the clinical study of biomolecular interactions over the last two decades (Nguyen et al. 2015). As a result, biosensors based on SPR have become popular in drug development and basic research. SPR instruments can provide information on the ligand-receptor binding levels, specificity, kinetics, and affinity of the interaction, thermodynamics, or molecule concentration (Frostell et al. 2013).

SPR occurs when a photon of incident light strikes a metal surface (typically a gold surface) (Figure 7). At a certain angle of incidence, a portion of the light energy couples through the metal coating with the electrons in the metal surface layer, which then move due to the excitation band (Link and El-Sayed 2003). Noble metal nanoparticles have a strong absorption band in the visible region which is attributed to their size. The coherent oscillation of the conduction band electrons caused by the interacting electromagnetic field is the physical origin of light absorption by metal nanoparticles in the SPR instrument. The surface effects are significant when the conductor dimensions are reduced. For this reason, the optical properties of small metal nanoparticles are dominated by the collective oscillation of conduction electrons, and when the frequency of the photon resonates with the collective oscillation of the conduction band electrons, an absorption band results and is known as the SPR (Ghosh and Pal 2007).

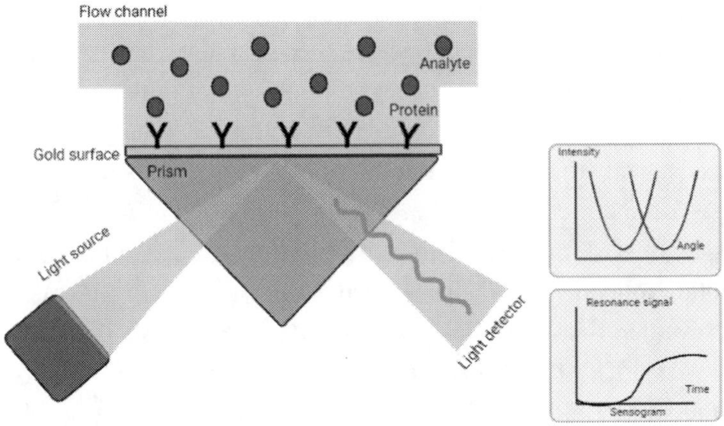

Figure 7. Schematic diagram of surface plasmon resonance.

In a commercial SPR biosensor configuration, incident light uses a high-reflective index glass prism in the Kretschmann geometry of the attenuated total reflection (ATR) method. The angle at which SPR is generated with a constant light source wavelength and a thin metal surface is determined by the substance's refractive index near the metal surface. As a result, a slight change in the reflective index of the sensing medium (for example, the interaction of the ligand with its receptor grafted to the sensor) will reduce the occurrence of SPR (Zhu and Gao 2019). Then, a detector is used to measure the change in the angle of the refractive index (Figure 7). A 0.1-degree change in angle is equivalent to a 1000-unit change in response units (RU), proportional to the surface mass (Wilson 2002).

Biosensor-based SPR has been widely utilized in the biomedical sector for the interaction of molecules based on affinity analysis, such as antibody-antigen (Hearty et al. 2012), epitope—mapping (Bhandari et al. 2019), and enzyme-substrate (Quintanilla-Villanueva et al. 2021) interactions. Another extension of SPR-based detection applications is its use in point mutation detection by combining SPR with other conventional techniques (Nguyen et al. 2015). Moreover, SPR could be used to study the interaction of phytochemicals with proteins related to virulence factors in bacteria measured by affecting optical indicators when a protein undergoes a structural change.

Even though SPR is a potent tool, few studies have evaluated the potential of phytochemicals as possible inhibitors of proteins related to the virulence of pathogenic bacteria. For example, Wang et al. (2021) conducted a study to search for inhibitors of cysteine transpeptidase sortase A (SrtA) of methicillin-resistant *Staphylococcus aureus* (MRSA). The authors evaluated the mechanism of binding of taxifolin, a flavonoid compound isolated from Chinese herbs, and SrtA using SPR assay. The results revealed that taxifolin binds directly to SrtA in a dose-dependent manner with a Kd of 1.65×10^{-4} mol/L.

In the same field, Shimamura et al. (2018) used SPR to evaluate the binding interaction of catechins with staphylococcal enterotoxin A toxin (SEA), that plays a crucial role in staphylococcal food poisoning. Epicatechin (100 μM) showed no interaction with the SEA immobilized on the chip, while epigallocatechin gallate (50 μM or more) strongly reacted with the toxin. These results suggested that the galloyl group directly interacts with SEA, preventing the antibodies from binding to the epitope of active toxin sites. A similar approach is the use of SPR to evaluate the interaction of potential inhibitors with other molecules that resulted in the inhibition of virulence factors. For example, Chang et al. (2019b) investigated the mechanism of action of different catechins to inhibit the activity of leukotoxin, a critical virulence factor in *Aggregatibacter actinomycetemcomitans*. In this study, the interaction between liposome cholesterol, toxin, and catechins was evaluated using SPR. These authors found that galloylated catechins affected the toxin conformation and its binding to cholesterol, which is crucial for the virulence mechanism.

As previously mentioned, few investigations have used the SPR to evaluate the interaction of phytochemicals with bacterial proteins; however, other synthetic molecules or obtained from different sources have been reported as potential virulence inhibitors ('D'Cunha et al. 2021, Borišek et al. 2018). In this sense, Chen et al. (2021) explored by SPR how the fusarielin M, a molecule isolated from marine-derived fungal strain *Fusarium gramnearum* interacts with the secreted protein tyrosine phosphatase B of *Mycobacterium tuberculosis*, a virulence factor responsible for intracellular endurance in host macrophages. The results showed that fusarielin M is bound to the bacterial protein with a Kd of 10.17 ± 0.90 μM, suggesting a moderate binding affinity. Similarly, the binding between 2-benzamidobenzoic acids and the PqsD enzyme of *P. aeruginosa* was assessed by SPR (Weidel et al. 2013). This enzyme catalyzes the biosynthesis of 2-heptyl-4-quinolone (HHQ), which is the precursor of the 2-heptyl-3-hydroxy-4 (1H)-quinolone (PQS) signal molecule involved in *P. aeruginosa quorum-sensing*, controlling some virulence factors. The SPR assays demonstrated that modified 2-benzamidobenzoic acids bind in the anthraniloyl-CoA channel preventing the substrate from accessing the active site. These studies can serve as the basis for future research in discovering new natural drugs from plants.

Nuclear Magnetic Resonance (NMR)

Over the last few decades, NMR spectroscopy has grown in popularity as a powerful tool for studying protein-ligand binding affinities in the nanomolar – millimolar range, as well as its ability to shed light on the protein conformational changes that may occur when they bind to their ligands (Maity et al. 2019). The phenomenon of NMR spectroscopy is that the nuclei of certain atoms can orient their nuclear spin by applying a magnetic field. Different orientations would have associated energies and could produce transitions when receiving radiation pulses (Deleanu and Paré 1997). These changes are represented in a diagram relating to the signal intensity vs. applied frequency. The resonance frequency of the nuclei depends on the chemical species, the type of nucleus, the intensity of the used magnetic field, and the chemical environment that surrounds the nuclei. This last factor can induce small changes in the signal frequency; this effect is known as chemical shift (Fritzsching et al. 2013).

The advantages of using NMR are numerous concerning its limitations. It allows studying the interactions directly in different media, usually in an aqueous solution, which is a great advantage since it is the only technique that enables the identification of the structure under these conditions (Sugiki et al. 2018). It also allows, in certain situations, to detect various conformations of the same compound and can detect and quantify interactions without knowing the protein function to which the drug will bind (Valente et al. 2006). One of the limitations of this method is that it can only be used for small molecules or complexes of these with macromolecules of molecular weight maximum of 30 kDa. Furthermore, the molecule's conformation is highly dependent on the environment, so a slight change in the environment can lead to significant differences in conformation (Everett 2007).

NMR spectroscopy is well suited for identifying and characterizing molecules that can be ideally placed between "target to tractable hit" and "tractable hit to a candidate." Chemical exchange, which occurs when a nucleus exchange between environments with structural characteristics that cause a difference in the local magnetic environment, can be used to observe dynamic events, and this step may be slow, intermediate, or fast depending on the time frame or frequencies (Cavanagh et al. 1995). These events can be intra-molecular or inter-molecular processes, with the former including protein side-chain motions, nucleic acid helix–coil transitions, protein unfolding, conformational equilibria, and tautomerization, and the latter including ligand binding to macromolecules, protonation/ deprotonation equilibria of ionizable groups (Campos-Olivas 2011, Bernetti et al. 2017).

One of the most exciting applications of NMR is detecting complexes of proteins with other biomolecules under physiological conditions at atomic resolution, even if these interactions are weak or transient. In this way, NMR has emerged as the only technique capable of studying, under these conditions, interactions of proteins with small ligands with an efficient application in rational drug design (Shi and Zhang 2021).

Furthermore, the results obtained by NMR made it possible to detect the existence of molecular recognition between two species, identify binding sites, and even the bioactive conformations of both species (without the need to know, in many cases, the structure of both molecules) (Unione et al. 2014, del Carmen Fernández-Alonso et al. 2012).

NMR is an attractive approach to study the interactions of phytochemicals and target proteins in the virulence process of pathogenic bacteria. However, very few studies have explored this option. Gradišar et al. (2007) determined the specific binding of epigallocatechin gallate to the N-terminal 24 kDa fragment of gyrase B by employing heteronuclear two-dimensional NMR spectroscopy. DNA gyrase is an excellent target for antibacterial agents since it is an essential enzyme in bacterial replication. The epigallocatechin gallate caused the amide groups in gyrase B to display a chemical shift perturbation proving information about the protein-ligand interaction in the formed complex.

In the same research area but using synthetic compounds, Hou et al. (2018) searched in the natural product-based clinical drug libraries a new class of SrtA inhibitors based in a quinone skeleton. SrtA inhibition causes the reduction in bacterial virulence and biofilm formation in *S. aureus*; therefore, it is an excellent target for anti-virulence strategies. NMR analysis in this study revealed that a quinone skeleton compound interacted with SrtA, acting as an irreversible inhibitor that covalently modifies its active site Cys184 (Hou et al. 2018). Similarly, Weidel et al. (2013) used the saturation transfer difference (STD) NMR technique to elucidate the binding mode of synthetic sulfonamide substituted 2-benzamidobenzoic acids to PqsD, an essential enzyme for the signal molecule formation in the *quorum-sensing* system of *P. aeruginosa*. STD NMR is a valuable tool for defining binding epitopes on small-molecule ligands with accuracy. The results of this study suggested that inhibitors could bind to the anthraniloyl-CoA channel of PqsD blocking its access to the catalytic site. This effect avoids the condensation of anthraniloyl-CoA with β-ketodecanoic acid to form HHQ, the signal molecule in the *quorum-sensing* system.

Methods to Evaluate Conformational Changes in Target Proteins

Circular Dichroism

Circular dichroism (CD) is an excellent tool for rapidly determining protein's secondary structure and folding properties. CD compares the magnitudes of absorption or scattering of a plane-polarized ray of light made up of two circularly-polarized components, one on the right and the other on the left (Kelly et al. 2005). These components are of the same amplitude, and each component interacts differently with the chiral centers of the studied molecules. The radiation interaction with the sample induces a differential phase shift and change in magnitude in both circularly polarized light components, causing a rotation of the polarization plane at an angle, and the distortion of this plane generates an ellipse (Miles et al. 2021). The dichroism spectra in the far ultraviolet region are mainly due to the amide bonds that link amino acid residues together. The asymmetry of chromophore or chiral groups is due to the spatial arrangement of the main chain of the protein. Therefore, the circular dichroism signals can be interpreted in terms of the content of secondary structures, that is, the percentage of residues found in some structural conformation (α-helix, β-sheet, turns, and other structural types).

Bacterial proteins are biological macromolecules with many chromophore groups present along the polypeptide chain (amino acids), which provide valuable information on conformation, stability, and function (Berova et al. 2000). The CD spectra for proteins are carried out in two regions of the visible ultraviolet from 180 to 240 nm and from 260 to 320 nm, the first ones are related to the secondary structure (the content of α-helices and β-sheets) while the latter to the tertiary structure and the behavior of aromatic amino acids. Both spectra allow evaluating the conformational changes of proteins induced by the binding of ligands.

The CD has proved highly useful for evidencing conformational changes caused by the binding of drugs or natural compounds on proteins.

This technique is a quick and simple spectroscopic assay that only requires a soluble protein without extensive sample preparation (Siligardi et al. 2014). The analysis of a CD spectrum provides an estimation of the number of secondary structures that make up the macromolecule or the polarity of the environment in which some aromatic amino acids are found. CD has the advantage of being a rapid method, which requires relatively small amounts of the sample under study, and which is very versatile since experimental conditions (pH, temperature, solutes) can be varied within relatively wide intervals. Due to these characteristics, the CD has been, together with fluorescence spectroscopy, the most widely used method in studying the stability of proteins and other macromolecules.

CD has been used as an effective tool to analyze the conformational changes of bacterial proteins related to virulence induced by the interaction with natural antibacterial and anti-virulence compounds (Chang et al. 2019a). These investigations aimed to determine the mode of action of these compounds and support the design of potential natural treatments. With this approach, it has been demonstrated that flavonoid compounds like myricetin interacts with α-hemolysin of *S. aureus,* reducing heptamer formation altering its secondary structure. These results were related to the attenuation of biofilm development and bacteria pathogenicity (Wang et al. 2020). On the other hand, it has been reported that glucovanillin, a phenolic compound, formed a complex with *Acinetobacter radioresistens* lipase reducing their alpha helix content from 31.3% to 3.1% and increasing beta-sheets levels (19.9–23.9%), causing conformational changes that modified the bacterial enzymatic stability. Furthermore, different phenolic compounds such as catechin, catechin gallate, epicatechin, epicatechin gallate, epigallocatechin, epigallocatechin gallate, and gallocatechin gallate interacted with *A. actinomycetemcomitans* leukotoxin, modifying its structure and avoiding the binding of the toxin to the host cell (Chang et al. 2019a). These studies show the valuable information obtained from CD experiments that can help elucidate the mode of action of phytochemicals with anti-virulence properties.

Ultraviolet and Visible Light Spectroscopy

UV-visible (UV-Vis) spectroscopy is a simple and valuable tool to study ligand-protein interactions (Alam et al. 2015). This technique is complementary to other assays such as fluorescence spectroscopy, circular dichroism, dynamic light scattering, and differential scanning calorimetry. UV-Vis spectroscopy fundament its use in the molecular ability to absorb radiation, including the visible UV spectrum. In spectroscopy, light is not only applied to the visible form of electromagnetic radiation but also to UV and IR forms, which are invisible. In absorbance spectrophotometry, the UV (near UV, 195-400 nm) and the visible (400-780 nm) regions are used (Penner 2017). The advantage of this method is that they operate with a small number of effective parameters. UV-vis spectrophotometry leads the evaluation of structural dynamics of local sites of large molecules (Demchenko 2013). The measurement of the absorbance of light by biomolecules is carried out in devices called spectrophotometers. This equipment consists of deuterium and tungsten lamps, a monochromator for the selection of radiation of a particular wavelength, a compartment where a container is housed transparent (quartz, glass, or plastic), a light detector, and an amplifier that converts light into electrical signals with a data reading system (Penner 2017).

The application of UV-Vis on protein-ligand interactions studies is based on the ability of proteins to absorb light energy (approximately 280 nm) and store it as internal energy. When a molecule absorbs light (considered energy), there is a jump from a basal or fundamental energy state to a higher energy state (Wang, Sun, Pu, Wei, et al. 2017). Only the energy that allows the jump to the excited state will be absorbed. Each biomolecule has a series of excited states (or bands) that distinguishes it from other molecules. Therefore, the absorption that a protein without conformational changes presents at different wavelengths constitutes a sign of its identity. Finally, the protein in excited form releases the absorbed energy to the ground energy state. Changes in protein conformation and dissociation and protein denaturation may lead to the change in one or more aromatic amino acid residues (Wang, Sun, Pu, and Wei 2017).

This technique is an analytical method that uses the mathematical transformation of a normal spectral curve into a derivative to extract qualitative and quantitative information from overlapping bands of the analytes. Changes in protein conformation can be evidenced by comparing the changes in the absorption spectra of the protein in the presence of an increasing ligand concentration. The principal information on chromophore interaction with the environment is provided by spectral shifts that may proceed in different directions. The term "blue shift" is synonymous with the shortwave shift (hypsochromic shift), while the "red shift" denotes the longwave shift (bathochromic shift) (Demchenko 2013). Blue shifts are related to a change of π–$\pi*$ transition brought about by changes in the conformational state of the protein and indicated an unfolding of the tertiary structure. The collected data from UV-Vis experiments can be used to calculate a foldedness ratio to investigate unfolding, refolding, disulfide bonds, stability, buffer excipients, and even protein-protein and protein-ligand interactions (Biter et al. 2019). The wavelengths of radiation and absorption efficiency that a molecule can present depend on the atomic structure and the conditions of the environment (pH, temperature, ionic forces, dielectric constant) (Verma and Mishra 2018).

A few studies applied this technique to study the conformational changes of proteins related to virulence upon ligand binding. Chaturvedi et al. (2015) studied the interaction of the terpene limonene with the model protein Bovine Serum Albumin (BSA) using spectroscopic methods, including UV-Vis spectroscopy. The V-visible spectra showed a change in the peaks within the aromatic region, indicating hydrophobic interactions with aromatic residues in the protein. Aromatic amino acids contribute to bands in the range of 255–300 nm. After adding limonene, the maximal absorption peak and the absorption intensity of BSA were increased, evidencing conformational changes in BSA due to its interaction with limonene (Chaturvedi et al. 2015). In addition, CD revealed an increase in α-helical contents probably on the cost of random coils or/and β-sheets of BSA.

In other studies, UV-Vis spectroscopy has been used to verify the integrity of protein structure upon binding with chemical compounds. Yang et al. (2013) used UV-Vis spectroscopy to study the interaction of the dye chrysoidine with Bovine Liver Catalase (BLC). No detectable change was observed for the BLC absorption spectrum at around 280 nm in the presence of chrysoidine, indicating that no profound conformational change of BLC occurs upon chrysoidine binding (Yang et al. 2013). On the other hand, Vignesh et al. (2012) used the UV–Vis absorption spectra of BSA obtained in the presence and absence of polymer–copper(II)bipyridine/phenanthroline to evidence the formation of a ground-state complex between BSA and the polymer–copper(II) complex. Although those studies performed by UV light absorption spectroscopy do not describe the whole protein conformation, they can be applied to analyze some properties of the conformational state of virulence proteins to reveal the dynamics at the level of interacting atoms groups (Demchenko 2013). The advantage of these methods is the simplicity of performance and non-destructiveness (Demchenko 2013).

Fluorescence Spectroscopy

Fluorescence spectroscopy is another biophysical tool that can investigate the conformational changes of proteins upon ligand binding. Fluorescence is light emission by excited atoms and molecules by absorbing light or electromagnetic radiation (Hellmann and Schneider 2019). When excited atoms or molecules relax to the ground state, they release excess energy in the form of photons. This fluorescence emission, which takes place in millionths of a second, induces modifications in the properties of fluorescent compounds. Fluorescence spectrometry or fluorometry is a type of spectroscopy that allows the analysis and measurement of the fluorescence pattern of a sample.

Fluorimeters are devices that make it possible to measure the fluorescence parameters, such as the intensity and distribution of wavelengths of the emission spectrum after excitation by a high-energy monochromatic light source. Filters and monochromators can be used in Fluorimeters (Singh et al. 2021). By a system of monochromators, it is possible to select the range of wavelengths of interest. The most common type of monochromator employs a diffraction grating, which allows collimated light entering the grating to exit at a different angle depending on the wavelength that has been selected. The addition of polarization filters is possible for anisotropy studies.

The sample to be analyzed is placed in a compartment, receiving both the direct light that passes through the glass and that emitted; then the detection system collects the absorbed or emitted light. To study conformational changes of proteins upon ligand binding, variants of this technique, such as synchronous fluorescence spectroscopy or three-dimensional fluorescence spectroscopic analysis, could be used (Chaturvedi et al. 2015). In both methods, shifts in the excitation and emission spectra could be related to conformational changes of the studied proteins in the presence of natural ligands or inhibitors (Chaturvedi et al. 2015).

Very few studies have explored the conformational changes of bacterial virulence protein when interacting with phytochemicals. However, this approach has been considered for other proteins, which could serve as background for future research in bacterial antivirulence therapy. For example, Moreno-Córdova et al. (2020) employed fluorescence quenching experiments and suggested that the binding of tannic acid and penta-Ogalloyl-β-D-glucose to the pancreatic lipase caused significant conformational changes to its tertiary structure, which may be correlated with their inhibitory effect and their potential use as inhibitors to treat dyslipidemias and obesity. Similarly, intrinsic protein fluorescence has been used to study the binding properties of whey proteins with caffeic acid and (-)-epigallocatechin gallate, as well as tertiary structure changes (Pessato et al. 2018). Fluorescence is a helpful technique for studying changes in the protein conformation and internal electrostatics in the presence of inhibitors, evidenced by changes in emission intensity (fluorescence quenching) (Moreno-Córdova et al. 2020).

Conclusion and Future Trends

Using this set of techniques can be a powerful tool to validate the sites of action of phytochemicals with anti-virulence properties and help elucidate the mechanism of action. These techniques can help describe the interaction of phytochemicals with pathogenic bacteria's cell membrane and their intercellular targets that can be correlated with phenotypic changes observed in bacteria treated with sublethal doses of these compounds. This information is of great importance since it allows phytochemicals to advance to the next stages of development and thus increases the availability of natural compounds capable of attenuating the virulence of pathogenic bacteria without exerting selective pressure. In addition, biophysical techniques have been refined, and the standardized protocols to obtain valuable information from multiple compounds at once have been automatized. The information collected can serve as the bio-inspiration for new therapeutic compounds. It is expected that the new generation of biophysical tools allows the *in-vivo* evaluation of the interaction of phytochemicals with pathogenic bacteria to help to fight the spread of antibiotic resistance.

References

Aitipamula, S., and Vangala, V. R. (2017). X-ray crystallography and its role in understanding physicochemical properties of pharmaceutical cocrystals. *Journal of the Indian Institute of Science* 97:2:227–243.

Alam, P., Chaturvedi, S. K., Anwar, T., Siddiqi, M. K., Ajmal, M. R., Badr, G., Mahmoud, M. H., and Hasan Khan, R. (2015). Biophysical and molecular docking insight into the interaction of cytosine β-D arabinofuranoside with human serum albumin. *Journal of Luminescence* 164:123-130.

Ameh, E. S. (2019). A review of basic crystallography and x-ray diffraction applications. *International Journal of Advanced Manufacturing Technology* 105 (7/8):3289-3302.

Archer, W. R., and Schulz, M. D. (2020). Isothermal titration calorimetry: practical approaches and current applications in soft matter. *Soft Matter* 16 (38):8760-8774.

Baranauskiene, L., Kuo, T.-C., Chen, W.-Y., and Matulis, D. (2019). Isothermal titration calorimetry for characterization of recombinant proteins. *Current Opinion in Biotechnology* 55:9-15.

Beckham, K. S. H., and Roe, A. J. (2014). From screen to target: insights and approaches for the development of anti-virulence compounds. *Frontiers in Cellular and Infection Microbiology* 4 (139).

Bernal-Mercado, A. T., Vazquez-Armenta, F. J., Tapia-Rodriguez, M. R., Islas-Osuna, M. A., Mata-Haro, V., Gonzalez-Aguilar, G. A., Lopez-Zavala, A. A., and Ayala-Zavala, J. F. (2018). Comparison of single and combined use of catechin, protocatechuic, and vanillic acids as antioxidant and antibacterial agents against uropathogenic *Escherichia coli* at planktonic and biofilm levels. *Molecules* 23 (11):2813.

Bernal-Mercado, A. T., Gutierrez-Pacheco, M. M., Encinas-Basurto, D., Mata-Haro, V., Lopez-Zavala, A. A., Islas-Osuna, M. A., Gonzalez-Aguilar, G. A., and Ayala-Zavala, J. F. (2020). Synergistic mode of action of catechin, vanillic and protocatechuic acids to inhibit the adhesion of uropathogenic *Escherichia coli* on silicone surfaces. *Journal of Applied Microbiology* 128 (2):387-400.

Bernetti, M., Cavalli, A., and Mollica, L. (2017). Protein–ligand (un) binding kinetics as a new paradigm for drug discovery at the crossroad between experiments and modelling. *Med Chem Comm* 8 (3):534-550.

Berova, N., Nakanishi, K., and Woody, R. W. (2000). *Circular dichroism: principles and applications*. 2 ed: John Wiley & Sons.

Bhandari, D., Chen, F. C., Hamal, S., and Bridgman, R. C. (2019). Kinetic analysis and epitope mapping of monoclonal antibodies to *Salmonella*

Typhimurium flagellin using a surface plasmon resonance biosensor. *Antibodies* (1):22.

Biter, A. B., Pollet, J., Chen, W.-H., Strych, U., Hotez, P. J., and Bottazzi, M. E. (2019). A method to probe protein structure from UV absorbance spectra. *Analytical Biochemistry* 587:113450.

Boisselier, É., Demers, É., Cantin, L., and Salesse, C. (2017). How to gather useful and valuable information from protein binding measurements using Langmuir lipid monolayers. *Advances in Colloid and Interface Science* 243:60-76.

Borges, A., Ferreira, C., Saavedra, M. J., and Simões, M. (2013). Antibacterial activity and mode of action of ferulic and gallic acids against pathogenic bacteria. *Microbial Drug Resistance* 19 (4):256-265.

Borišek, J., Pintar, S., Ogrizek, M., Grdadolnik, S. G., Hodnik, V., Turk, D., Perdih, A., and Novič, M. (2018). Discovery of (phenylureido) piperidinyl benzamides as prospective inhibitors of bacterial autolysin E from *Staphylococcus aureus*. *Journal of Enzyme Inhibition and Medicinal Chemistry* 33 (1):1239-1247.

Camejo, A., Carvalho, F., Reis, O., Leitão, E., Sousa, S., and Cabanes, D. (2011). The arsenal of virulence factors deployed by *Listeria monocytogenes* to promote its cell infection cycle. *Virulence* 2 (5):379-394.

Campbell, I. (2012). *Biophysical techniques*: Oxford University Press.

Campos-Olivas, R. (2011). NMR screening and hit validation in fragment based drug discovery. *Current topics in medicinal chemistry* 11 (1):43-67.

Cavanagh, J., Fairbrother, W. J., Palmer III, A. G., and Skelton, N. J. (1995). *Protein NMR spectroscopy: principles and practice*: Elsevier.

Chaires, J. B. (2008). Calorimetry and thermodynamics in drug design. *Annual review of biophysics* 37:135-51.

Chang, E. H., Huang, J., Lin, Z., and Brown, A. C. (2019a). Catechin-mediated restructuring of a bacterial toxin inhibits activity. *Biochimica et Biophysica Acta (BBA)-General Subjects* 1863 (1):191-198.

Chang, E. H., Huang, J., Lin, Z., and Brown, A. C. (2019b). Catechin-mediated restructuring of a bacterial toxin inhibits activity. *Biochimica et Biophysica Acta -General Subjects* 1863 (1):191-198.

Chaturvedi, S. K., Ahmad, E., Khan, J. M., Alam, P., Ishtikhar, M., and Khan, R. H. (2015). Elucidating the interaction of limonene with bovine serum albumin: a multi-technique approach. *Molecular BioSystems* 11 (1):307-316.

Chen, D., Liu, L., Lu, Y., and Chen, S. (2021). Identification of fusarielin M as a novel inhibitor of *Mycobacterium tuberculosis* protein tyrosine phosphatase B (MptpB). *Bioorganic Chemistry* 106:104495.

Choudhury, P. R., Talukdar, A. D., Nath, D., Saha, P., and Nath, R. (2020). Traditional Folk Medicine and Drug Discovery: Prospects and Outcome. In *Advances in Pharmaceutical Biotechnology: Recent Progress and Future Applications*, edited by Jayanta Kumar Patra, Amritesh C. Shukla and Gitishree Das, 3-13. Singapore: Springer Singapore.

Corral-Lugo, A., Daddaoua, A., Ortega, A., Espinosa-Urgel, M., and Krell, T. (2016). Rosmarinic acid is a homoserine lactone mimic produced by plants that activates a bacterial quorum-sensing regulator. *Science Signaling* 9 (409):ra1-ra1.

D'Cunha, N., Moniruzzaman, M., Haynes, K., Malloci, G., Cooper, C. J., Margiotta, E., Vargiu, A. V., Uddin, M. R., Leus, I. V., and Cao, F. (2021). Mechanistic duality of bacterial efflux substrates and inhibitors: Example of simple substituted cinnamoyl and naphthyl amides. *ACS Infectious Diseases* 7 (9):2650–2665.

Dai, L., Li, Z., Chen, D., Jia, L., Guo, J., Zhao, T., and Nordlund, P. (2020). Target identification and validation of natural products with label-free methodology: A critical review from 2005 to 2020. *Pharmacology & Therapeutics* 216:107690.

de Carvalho, A. C., Girola, N., de Figueiredo, C. R., Machado, A. C., de Medeiros, L. S., Guadagnin, R. C., Caseli, L., and Veiga, T. A. (2019). Understanding the cytotoxic effects of new isovanillin derivatives through phospholipid Langmuir monolayers. *Bioorganic Chemistry* 83:205-213.

del Carmen Fernández-Alonso, M., Díaz, D., Alvaro Berbis, M., Marcelo, F., and Jimenez-Barbero, J. (2012). Protein-carbohydrate interactions studied by NMR: from molecular recognition to drug design. *Current Protein and Peptide Science* 13 (8):816-830.

Deleanu, C., and Paré, J. J. (1997). Nuclear magnetic resonance spectroscopy (NMR): Principles and applications. In *Techniques and Instrumentation in Analytical Chemistry*, 179-237. Elsevier.

Demchenko, A. P. (2013). *Ultraviolet Spectroscopy of Proteins*: Springer Berlin Heidelberg.

Díaz-Nuñez, J. L., García-Contreras, R., and Castillo-Juárez, I. (2021). The New Antibacterial Properties of the Plants: Quo vadis Studies of Anti-virulence Phytochemicals? *Frontiers in Microbiology* 12 (1054).

Elderdfi, M., and Sikorski, A. F. (2018). Langmuir-monolayer methodologies for characterizing protein-lipid interactions. *Chemistry and Physics of Lipids* 212:61-72.

Everett, J. R. (2007). Drug discovery and development: the role of NMR. *eMagRes*:137-150.

Ferreira, J. o. V. N., Capello, T. M., Siqueira, L. J., Lago, J. o. H. G., and Caseli, L. (2016). Mechanism of action of thymol on cell membranes investigated through lipid Langmuir monolayers at the air–water interface and molecular simulation. *Langmuir* 32 (13):3234-3241.

Folmer, R. H. A. (2016). Integrating biophysics with HTS-driven drug discovery projects. *Drug Discovery Today* 21 (3):491-498.

Franzoni, G., Trivellini, A., Bulgari, R., Cocetta, G., and Ferrante, A. (2019). Chapter 10 - Bioactive Molecules as Regulatory Signals in Plant Responses to Abiotic Stresses. In *Plant Signaling Molecules*, edited by M. Iqbal R. Khan, Palakolanu Sudhakar Reddy, Antonio Ferrante and Nafees A. Khan, 169-182. Woodhead Publishing.

Fritzsching, K., Yang, Y., Schmidt-Rohr, K., and Hong, M. (2013). Practical use of chemical shift databases for protein solid-state NMR: 2D chemical shift maps and amino-acid assignment with secondary-structure information. *Journal of Biomolecular NMR* 56 (2):155-167.

Frostell, Å., Vinterbäck, L., and Sjöbom, H. (2013). Protein–Ligand Interactions Using SPR Systems. In *Protein-Ligand Interactions*, 139-165. Springer.

Genick, C. C., and Wright, S. K. (2017). Biophysics: for HTS hit validation, chemical lead optimization, and beyond. *Expert Opinion on Drug Discovery* 12 (9):897-907.

Ghosh, S. K., and Pal, T. (2007). Interparticle coupling effect on the surface plasmon resonance of gold nanoparticles: from theory to applications. *Chemical Reviews* 107 (11):4797-4862.

Gradišar, H., Pristovšek, P., Plaper, A., and Jerala, R. (2007). Green tea catechins inhibit bacterial DNA gyrase by interaction with its ATP binding site. *Journal of Medicinal Chemistry* 50 (2):264-271.

Grundner, C., Perrin, D., Hooft van Huijsduijnen, R., Swinnen, D., Gonzalez, J., Gee, C. L., Wells, T. N., and Alber, T. (2007). Structural Basis for Selective Inhibition of *Mycobacterium tuberculosis* Protein Tyrosine Phosphatase PtpB. *Structure* 15 (4):499-509.

Hac-Wydro, K., and Dynarowicz-Latka, P. 2008. *Biomedical applications of the Langmuir monolayer technique*. Annales Universitatis Mariae Curie-Sklodowska.

Hearty, S., Leonard, P., and O'Kennedy, R. (2012). Measuring antibody–antigen binding kinetics using surface plasmon resonance. In *Antibody Engineering*, 411-442. Springer.

Hellmann, N., and Schneider, D. (2019). Hands on: using tryptophan fluorescence spectroscopy to study protein structure. In *Protein Supersecondary Structures*, 379-401. Springer.

Hodnik, V., and Anderluh, G. (2013). Surface plasmon resonance for measuring interactions of proteins with lipid membranes. In *Lipid-Protein Interactions*, 23-36. Springer.

Hou, X., Wang, M., Wen, Y., Ni, T., Guan, X., Lan, L., Zhang, N., Zhang, A., and Yang, C.-G. (2018). Quinone skeleton as a new class of irreversible inhibitors against *Staphylococcus aureus* sortase A. *Bioorganic and Medicinal Chemistry Letters* 28 (10):1864-1869.

Ionescu, D., Margină, D., Ilie, M., Iftime, A., and Ganea, C. (2013). Quercetin and epigallocatechin-3-gallate effect on the anisotropy of model membranes with cholesterol. *Food and Chemical Toxicology* 61:94-100.

Jacquemard, C., Drwal, M. N., Desaphy, J., and Kellenberger, E. (2019). Binding mode information improves fragment docking. *Journal of Cheminformatics* 11 (1):24.

Kalas, V., Hibbing, M. E., Maddirala, A. R., Chugani, R., Pinkner, J. S., Mydock-McGrane, L. K., Conover, M. S., Janetka, J. W., and Hultgren, S. J. (2018). Structure-based discovery of glycomimetic FmlH ligands as inhibitors of bacterial adhesion during urinary tract infection. *Proceedings of the National Academy of Sciences* 115 (12):E2819-E2828.

Kamal, A. A. M., Maurer, C. K., Allegretta, G., Haupenthal, J., Empting, M., and Hartmann, R. W. (2018). Quorum Sensing Inhibitors as Pathoblockers for *Pseudomonas aeruginosa* Infections: A New Concept in Anti-Infective Drug Discovery. In *Antibacterials: Volume II*, edited by Jed F. Fisher, Shahriar Mobashery and Marvin J. Miller, 185-210. Cham: Springer International Publishing.

Kasahara, K., and Kinoshita, K. (2016). Landscape of protein–small ligand binding modes. *Protein Science* 25 (9):1659-1671.

Keily, H. J., and Hume, D. N. (1956). Thermometric Titration Curves. *Analytical Chemistry* 28 (8):1294-1297.

Kelly, S. M., Jess, T. J., and Price, N. C. (2005). How to study proteins by circular dichroism. *Biochimica et Biophysica Acta (BBA) - Proteins and Proteomics* 1751 (2):119-139.

Lea, W. A., and Simeonov, A. (2011). Fluorescence polarization assays in small molecule screening. *Expert Opinion on Drug Discovery* 6 (1):17-32.

Li, H., Chen, Y., Zhang, B., Niu, X., Song, M., Luo, Z., Lu, G., Liu, B., Zhao, X., Wang, J., and Deng, X. (2016). Inhibition of sortase A by chalcone prevents *Listeria monocytogenes* infection. *Biochemical Pharmacology* 106:19-29.

Liang, Y. (2008). Applications of isothermal titration calorimetry in protein science. *Acta Biochimica et Biophysica Sinica* 40 (7):565-576.

Link, S., and El-Sayed, M. A. (2003). Optical spectroscopy of surface plasmons in metal nanoparticles. In *Semiconductor and Metal Nanocrystals*, 415-444. CRC Press.

Londoño-Londoño, J., De Lima, V. R., Jaramillo, C., and Creczynski-Pasa, T. (2010). Hesperidin and hesperetin membrane interaction: understanding the role of 7-O-glycoside moiety in flavonoids. *Archives of Biochemistry and Biophysics* 499 (1-2):6-16.

Luna-Solorza, J. M., Vazquez-Armenta, F. J., Bernal-Mercado, A. T., Gutierrez-Pacheco, M. M., Nazzaro, F., and Ayala-Zavala, J. F. (2020). Phytochemical Compounds Targeting the Quorum Sensing System as a Tool to Reduce the Virulence Factors of Food Pathogenic Bacteria. In *Trends in Quorum Sensing and Quorum Quenching*, edited by Jamuna A Bai V. Ravishankar Rai, 257-276. Boca Raton: CRC Press.

Maget-Dana, R. (1999). The monolayer technique: a potent tool for studying the interfacial properties of antimicrobial and membrane-lytic peptides and their interactions with lipid membranes. *Biochimica et Biophysica Acta -Biomembranes* 1462 (1-2):109-140.

Maity, S., Gundampati, R. K., and Suresh Kumar, T. K. (2019). NMR methods to characterize protein-ligand interactions. *Natural Product Communications* 14 (5):1934578X19849296.

Matthews, H., Hanison, J., and Nirmalan, N. (2016). "Omics"- Informed Drug and Biomarker Discovery: Opportunities, Challenges and Future Perspectives. *Proteomes* 4 (3):28.

Maveyraud, L., and Mourey, L. (2020). Protein X-ray crystallography and drug discovery. *Molecucles* 25 (5):1030.

Menéndez, M. (2021). Isothermal Titration Calorimetry: Principles and Applications. In *eLS*, edited by Ltd (Ed.). John Wiley & Sons, 113-127. John Wiley & Sons.

Miles, A., Janes, R. W., and Wallace, B. A. (2021). Tools and methods for circular dichroism spectroscopy of proteins: a tutorial review. *Chemical Society Reviews* 50:8400-8413.

Moreno-Córdova, E. N., Arvizu-Flores, A. A., Valenzuela-Soto, E. M., García-Orozco, K. D., Wall-Medrano, A., Alvarez-Parrilla, E., Ayala-

Zavala, J. F., Dominguez-Avila A, and González-Aguilar, G. A. (2020). Gallotannins are uncompetitive inhibitors of pancreatic lipase activity. *Biophysical Chemistry* 264, 106409.

Mykytczuk, N., Trevors, J., Leduc, L., and Ferroni, G. (2007). Fluorescence polarization in studies of bacterial cytoplasmic membrane fluidity under environmental stress. *Progress in Biophysics and Molecular Biology* 95 (1-3):60-82.

Nakano, K., Chigira, T., Miyafusa, T., Nagatoishi, S., Caaveiro, J. M. M., and Tsumoto, K. (2015). Discovery and characterization of natural tropolones as inhibitors of the antibacterial target CapF from *Staphylococcus aureus*. *Scientific Reports* 5 (1):15337.

Nguyen, H. H., Park, J., Kang, S., and Kim, M. (2015). Surface plasmon resonance: a versatile technique for biosensor applications. *Sensors* 15 (5):10481-10510.

Nowotarska, S. W., Nowotarski, K. J., Friedman, M., and Situ, C. (2014). Effect of structure on the interactions between five natural antimicrobial compounds and phospholipids of bacterial cell membrane on model monolayers. *Molecules* 19 (6):7497-7515.

Omojate Godstime, C., Enwa Felix, O., Jewo Augustina, O., and Eze Christopher, O. (2014). Mechanisms of antimicrobial actions of phytochemicals against enteric pathogens–a review. *Journal of Pharmaceutical, Chemical and Biological Sciences* 2 (2):77-85.

Ooi, L. (2010). *Principles of X-ray Crystallography*: OUP Oxford.

Penner, M. H. (2017). Basic principles of spectroscopy. In *Food analysis*, 79-88. Springer.

Pessato, T. B., de Morais, F. P., de Carvalho, N. C., Figueira, A. C. M., Fernandes, L. G. R., Zollner, R. d. L., and Netto, F. M. (2018). Protein structure modification and allergenic properties of whey proteins upon interaction with tea and coffee phenolic compounds. *Journal of Functional Foods* 51, 121-129.

Peterson, J. W. (1996). Bacterial Pathogenesis. Edited by Baron S. 4th ed, *Medical Microbiology. 4th edition.* Galveston (TX): The University of Texas Medical Branch at Galveston.

Petty, M. C. (1996). *Langmuir-Blodgett films: an introduction*: Cambridge University Press.

Quintanilla-Villanueva, G. E., Luna-Moreno, D., Blanco-Gámez, E. A., Rodríguez-Delgado, J. M., Villarreal-Chiu, J. F., and Rodríguez-Delgado, M. M. (2021). A Novel Enzyme-Based SPR Strategy for Detection of the Antimicrobial Agent Chlorophene. *Biosensors* 11 (2):43.

Renaud, J. P., Chung, C. W., Danielson, U. H., Egner, U., Hennig, M., Hubbard, R. E., and Nar, H. (2016). Biophysics in drug discovery: impact, challenges and opportunities. *Nature Reviews Drug Discovery* 15 (10):679-698.

Roselin, L. S., Lin, M. S., Lin, P. H., Chang, Y., and Chen, W. Y. (2010). Recent trends and some applications of isothermal titration calorimetry in biotechnology. *Biotechnology Journal* 5 (1):85-98.

Selvaraj, S., Krishnaswamy, S., Devashya, V., Sethuraman, S., and Krishnan, U. M. (2015). Influence of membrane lipid composition on flavonoid–membrane interactions: Implications on their biological activity. *Progress in Lipid Research* 58:1-13.

Shi, L., and Zhang, N. (2021). Applications of Solution NMR in Drug Discovery. *Molecules* 26 (3):576.

Shimamura, Y., Utsumi, M., Hirai, C., Nakano, S., Ito, S., Tsuji, A., Ishii, T., Hosoya, T., Kan, T., and Ohashi, N. (2018). Binding of catechins to staphylococcal enterotoxin A. *Molecules* 23 (5):1125.

Siligardi, G., Hussain, R., Patching, S. G., and Phillips-Jones, M. K. (2014). Ligand-and drug-binding studies of membrane proteins revealed through circular dichroism spectroscopy. *Biochimica et Biophysica Acta (BBA) - Enzymology* 1838 (1):34-42.

Simoes, M., Bennett, R. N., and Rosa, E. A. (2009). Understanding antimicrobial activities of phytochemicals against multidrug resistant bacteria and biofilms. *Natural Product Reports* 26 (6):746-757.

Simpkin, V. L., Renwick, M. J., Kelly, R., and Mossialos, E. (2017). Incentivising innovation in antibiotic drug discovery and development: progress, challenges and next steps. *International Journal of Antibiotics* 70 (12):1087-1096.

Singh, D. P., Inamdar, S. R., Kumar, S., and Applications. (2021). Fluorescence Spectrometry. In *Modern Techniques of Spectroscopy: Basics, Instrumentation, and Applications*, 431. Springer Nature, 2021.

Smart, O. S., Horský, V., Gore, S., Svobodová Vařeková, R., Bendová, V., Kleywegt, G. J., and Velankar, S. (2018). Validation of ligands in macromolecular structures determined by X-ray crystallography. *Acta Crystallographica Section D* 74 (3):228-236.

Sommer, R., Hauck, D., Varrot, A., Wagner, S., Audfray, A., Prestel, A., Möller, H. M., Imberty, A., and Titz, A. (2015). Cinnamide Derivatives of d-Mannose as Inhibitors of the Bacterial Virulence Factor LecB from *Pseudomonas aeruginosa*. *Chemistry Open* 4 (6):756-767.

Stefaniu, C., Brezesinski, G., and Möhwald, H. (2014). Langmuir monolayers as models to study processes at membrane surfaces. *Advances in Colloid and Interface Science* 208:197-213.

Steinbach, A., Scheidig, A. J., and Klein, C. D. (2008). The unusual binding mode of cnicin to the antibacterial target enzyme MurA revealed by X-ray crystallography. *Journal of Medicinal Chemistry* 51 (16):5143-5147.

Storz, M. P., Brengel, C., Weidel, E., Hoffmann, M., Hollemeyer, K., Steinbach, A., Müller, R., Empting, M., and Hartmann, R. W. (2013). Biochemical and Biophysical Analysis of a Chiral PqsD Inhibitor Revealing Tight-binding Behavior and Enantiomers with Contrary Thermodynamic Signatures. *ACS Chemical Biology* 8 (12):2794-2801.

Sugiki, T., Furuita, K., Fujiwara, T., and Kojima, C. (2018). Current NMR techniques for structure-based drug discovery. *Molecules* 23 (1):148.

Sui, S. J. H., Fedynak, A., Hsiao, W. W., Langille, M. G., and Brinkman, F. S. (2009). The association of virulence factors with genomic islands. *PloS one* 4 (12):e8094.

Suresh, A., and Abraham, J. (2020). Phytochemicals and Their Role in Pharmaceuticals. In *Advances in Pharmaceutical Biotechnology: Recent Progress and Future Applications*, edited by Jayanta Kumar

Patra, Amritesh C. Shukla and Gitishree Das, 193-218. Singapore: Springer Singapore.

Tahrioui, A., Ortiz, S., Azuama, O. C., Bouffartigues, E., Benalia, N., Tortuel, D., Maillot, O., Chemat, S., Kritsanida, M., Feuilloley, M., Orange, N., Michel, S., Lesouhaitier, O., Cornelis, P., Grougnet, R., Boutefnouchet, S., and Chevalier, S. (2020). Membrane-Interactive Compounds From *Pistacia lentiscus* L. Thwart *Pseudomonas aeruginosa* Virulence. *Frontiers in Microbiology* 11 (1068).

Tapia-Rodriguez, M. R., Bernal-Mercado, A. T., Gutierrez-Pacheco, M. M., Vazquez-Armenta, F. J., Hernandez-Mendoza, A., Gonzalez-Aguilar, G. A., Martinez-Tellez, M. A., Nazzaro, F., and Ayala-Zavala, J. F. (2019). Virulence of *Pseudomonas aeruginosa* exposed to carvacrol: Alterations of the Quorum sensing at enzymatic and gene levels. *Journal of Cell Communitacion and Signaling* 13 (4):531-537.

Titz, A. (2014). Carbohydrate-Based Anti-Virulence Compounds Against Chronic *Pseudomonas aeruginosa* Infections with a Focus on Small Molecules. In *Carbohydrates as Drugs*, edited by Peter H. Seeberger and Christoph Rademacher, 169-186. Cham: Springer International Publishing.

Tiwari, K. B., Sen, S., Gatto, C., and Wilkinson, B. J. (2021). Fluorescence Polarization (FP) Assay for Measuring *Staphylococcus aureus* Membrane Fluidity. In *Staphylococcus aureus. Methods in Molecular Biology.*, edited by K. C. Rice, 55-68. New York: Springer.

Trevors, J. (2003). Fluorescent probes for bacterial cytoplasmic membrane research. *Journal of Biochemical and Biophysical Methods* 57 (2):87-103.

Tsuchiya, H. (2010). Structure-dependent membrane interaction of flavonoids associated with their bioactivity. *Food Chemistry* 120 (4):1089-1096.

Tsuchiya, H. (2015). Membrane interactions of phytochemicals as their molecular mechanism applicable to the discovery of drug leads from plants. *Molecules* 20 (10):18923-18966.

Tsuchiya, H., and Iinuma, M. (2000). Reduction of membrane fluidity by antibacterial sophoraflavanone G isolated from *Sophora exigua*. *Phytomedicine* 7 (2):161-165.

Tyson, B. C., McCurdy, W., and Bricker, C. (1961). Differential Thermometric Titrations and the Determination of Heats of Reaction. *International Journal of Analytical Chemistry* 33 (12):1640-1645.

Unione, L., Galante, S., Diaz, D., and Jiménez-Barbero, J. (2014). NMR and molecular recognition. The application of ligand-based NMR methods to monitor molecular interactions. *MedChemComm* 5 (9):1280-1289.

Vachali, P. P., Li, B., Besch, B. M., and Bernstein, P. S. (2016). Protein-flavonoid interaction studies by a taylor dispersion surface plasmon resonance (spr) technique: A novel method to assess biomolecular interactions. *Biosensors* 6 (1):6.

Valente, A., Miyamoto, C., and L Almeida, F. (2006). Implications of protein conformational diversity for binding and development of new biological active compounds. *Current Medicinal Chemistry* 13 (30):3697-3703.

Velázquez, M. M., Alejo, T., López-Díaz, D., Martín-García, B., and Merchán, M. D. (2016). Langmuir-Blodgett Methodology: A Versatile Technique to Build 2D Material Films. In *Two-dimensional Materials: Synthesis, Characterization and Potential Applications*, edited by Pramoda Kumar Nayak, 21. BoD – Books on Demand, 2016.

Verma, G., and Mishra, M. (2018). Development and optimization of UV-Vis spectroscopy-a review. *World Journal of Pharmaceutical Research* 7 (11):1170-1180.

Vignesh, G., Arunachalam, S., Vignesh, S., and James, R. A. (2012). BSA binding and antimicrobial studies of branched polyethyleneimine–copper(II)bipyridine/phenanthroline complexes. *Spectrochimica Acta Part A: Molecular and Biomolecular Spectroscopy* 96:108-116.

Wang, J., Liu, B., Teng, Z., Zhou, X., Wang, X., Zhang, B., Lu, G., Niu, X., Yang, Y., and Deng, X. (2017). *Phloretin Attenuates Listeria monocytogenes Virulence Both In vitro and In vivo by Simultaneously Targeting Listeriolysin O and Sortase A.* 7 (9).

Wang, K., Sun, D.-W., Pu, H., and Wei, Q. (2017). Principles and applications of spectroscopic techniques for evaluating food protein conformational changes: A review. *Trends in Food Science & Technology* 67:207-219.

Wang, L., Wang, G., Qu, H., Wang, K., Jing, S., Guan, S., Su, L., Li, Q., and Wang, D. (2021). Taxifolin, an inhibitor of sortase A, interferes with the adhesion of methicillin-resistant *Staphylococcal aureus*. *Frontier in Microbiology* 12:1876.

Wang, M., Zhao, L., Wu, H., Zhao, C., Gong, Q., and Yu, W. J. M. D. (2020). *Cladodionen Is a Potential Quorum Sensing Inhibitor Against Pseudomonas aeruginosa*. 18 (4):205.

Wilson, W. D. (2002). Analyzing biomolecular interactions. *Science* 295 (5562):2103-2105.

Wu, T., He, M., Zang, X., Zhou, Y., Qiu, T., Pan, S., and Xu, X. (2013). A structure–activity relationship study of flavonoids as inhibitors of *E. coli* by membrane interaction effect. *Biochimica et Biophysica Acta - Biomembranes* 1828 (11):2751-2756.

Yang, B., Hao, F., Li, J., Chen, D., and Liu, R. (2013). Binding of chrysoidine to catalase: Spectroscopy, isothermal titration calorimetry and molecular docking studies. *Journal of Photochemistry and Photobiology B: Biology* 128:35-42.

Zender, M., Klein, T., Henn, C., Kirsch, B., Maurer, C. K., Kail, D., Ritter, C., Dolezal, O., Steinbach, A., and Hartmann, R. W. (2013). Discovery and Biophysical Characterization of 2-Amino-oxadiazoles as Novel Antagonists of PqsR, an Important Regulator of *Pseudomonas aeruginosa* Virulence. *Journal of Medicinal Chemistry* 56 (17):6761-6774.

Zhu, X., and Gao, T. (2019). Spectrometry. In *Nano-Inspired Biosensors for Protein Assay with Clinical Applications*, 237-264. Elsevier.

In: Pathogenic Bacteria
Editor: Keith D. Watts
ISBN: 978-1-68507-422-7

Chapter 3

ANTIVIRULENCE MECHANISMS OF PLANT TERPENES AGAINST PATHOGENIC BACTERIA

Melvin R. Tapia-Rodriguez[1], F. Javier Vazquez-Armenta[2], Cristobal J. Gonzalez-Perez[3], Yessica Enciso-Martinez[3], Maria Gonzalez-Leyva[3] and J. Fernando Ayala-Zavala[3,*]

[1]Departamento de Biotecnología y Ciencias Alimentarias, Instituto Tecnologico de Sonora, Ciudad Obregon, Sonora, Mexico
[2]Departamento de Ciencias Químico Biológicas, Universidad de Sonora, Hermosillo, Sonora, Mexico
[3]Coordinacion de Tecnologia de Alimentos de Origen Vegetal, Centro de Investigacion en Alimentacion y Desarrollo, Hermosillo, Sonora, Mexico

[*] Corresponding Author's E-mail: jayala@ciad.mx.

Abstract

Pathogenic bacteria outbreaks cause international economic losses and public health issues during food production and clinical environments. New non-toxic and efficient treatments for disinfecting are highly demanded these days. In this context, terpenes are natural compounds widespread in plants and have attracted attention due to their antibacterial and anti-virulence potential. This chapter discusses the terpenes' anti-virulence mode of action, inhibition of biofilm, disruption of the cell wall and bacterial membrane, intercellular leakage, interruption of nucleic acids transcription, and interference with the quorum sensing signaling process. Overall, promising applications of terpenes compounds in the clinical and food industry as natural antibacterial treatments that effectively reduce pathogenicity and control resistant bacteria are evidenced.

Keywords: natural compounds, antibacterial, anti-virulence

Introduction

Pathogenic bacteria outbreaks are a global public health problem that causes economic losses (Wittry and Nicholas, 2020). Pathogenic bacteria possess different virulence factors, like motility, flagellum, and biofilm development, used to persist and resist disinfection and antibiotics (Luna-Solorza et al., 2020). The attached cells in biofilms are covered for different polymeric substances that interfere with antibiotics efficacy. In this context, it has been demonstrated that *quorum sensing* regulates virulence factors, including biofilm development, motility, and toxin production in *Pseudomonas aeruginosa* (Tapia-Rodriguez et al., 2017). *Quorum sensing* is a signaling process that controls genes expression related to virulence factors, and it is mediated by auto-inducer molecules recognized by protein receptors. Nowadays, it has been mentioned that interrupting this communication process can be a suitable target to reduce virulence factors and control these pathogens (Nazzaro et al., 2013).

Plant terpenes have been studied as an alternative to reduce pathogenic bacteria outbreaks (Gutierrez-Pacheco et al., 2019). In this context, carvacrol is found in several plants, especially in oregano, and it inhibits the growth of *Salmonella* Typhimurium, *Staphylococcus aureus* and *Listeria monocytogenes*. It affects bacterial membranes by promoting intercellular leakage that denatures protein structure and nucleic acid functions (Nostro et al., 2012). In addition, this terpene can affect intercellular communication of the biosensor strain *Chromobacterium violaceum*, reducing violacein synthesis; it is a pigment produced during the *quorum sensing* signals (Alvarez et al., 2014). In this perspective, this chapter explores the potential of terpenes compounds to exert their anti-virulence potential on pathogenic bacteria, affecting their biofilm formation, motility, toxin production, membrane, and cell wall damage, as well as interruption of *quorum sensing*.

EFFECT OF TERPENES ON BIOFILM FORMATION

Biofilms are surface-associated bacterial communities that differ from their planktonic counterparts in physiology, gene expression, and morphology (Borges et al., 2013). Attached cells in biofilms are surrounded by a matrix of extracellular polymeric substances (EPS) composed of polysaccharides, proteins, and DNA that strengthen the attachment to the surface and protect against environmental hazards (Hobley et al., 2015; Zeraik and Nitschke, 2012). Biofilm formation by pathogenic bacteria on medical devices and hospital environments is related to hospital-acquired infections and represents a serious threat to modern medicine (Floyd et al., 2017). Also, biofilm infections are the main cause of the failure of biomedical implants since all medical devices are susceptible to this phenomenon (Veerachamy et al., 2014). Additionally, contaminated medical devices can act as bacterial reservoirs promoting infections on the surrounding tissue.

In biofilms formed on medical devices, EPS provides protection against immune defenses and the action of antibiotics, making biofilm-associated infections complicated to treat (Veerachamy et al., 2014).

Some mechanisms of resistance of bacterial biofilms have been proposed including, i) poor penetration or inactivation of antibiotics in biofilm matrix, (ii) changes in outer membrane structure, (iii) an altered metabolic state of bacteria, (iv) the presence of persisted populations, (v) genetic adaptation, and (vi) resistance induced by antibiotics (Borges et al., 2013). For this reason, the understanding of the biofilm formation process of pathogenic bacteria could help develop new strategies for preventing biofilm-associated infections.

Biofilm formation is described in five consecutive steps: i) reversible adhesion, ii) irreversible adhesion, iii) microcolony formation, iv) maturation, and v) biofilm dispersion (Srey et al., 2013). Reversible adhesion is an initial weak interaction of bacteria, which have a surface-negative or -positive charge, and the substratum or abiotic surface. It involves balancing attractive van der Waals forces and repulsive and attractive electrostatic forces (AlAbbas et al., 2012). Once the bacteria overcome the repulsion forces of the surface and come into contact with it, phenotypic changes occur, including the synthesis of bacterial adhesins allowing the irreversible association between bacteria and surface (Verstraeten et al., 2008; Zeraik and Nitschke, 2012). In the next steps, microcolonies are developed, and biofilm maturation is characterized by increasing the EPS production, representing about 50 to 90% of total organic matter in mature bacterial biofilms. (Zhu et al., 2015).

The biofilm is an evolving process mediated by physicochemical interactions, bacterial motility, and bacterial communication. Since the best strategy for biofilm control is to prevent their development, the cellular functions of bacterial adhesion and biofilm formation are important targets (Donlan, 2002). In this context, several studies (Table 1) evidenced the antibiofilm properties of terpenes against pathogenic bacteria acting on different points, making them good candidates for anti-virulence therapy.

Table 1. Anti-biofilm properties of terpenes against pathogenic bacteria

Compound	Concentration	Bacteria	Adhesion or biofilm inhibition	Assay conditions	Mode of action	Reference
Carvacrol	0.0128%	*S. aureus*	71.7%	Polystyrene; 24 h; 37°C	Not evaluated	(Nostro et al., 2007)
		S. epidermis	42.9%			
	0.65 mM	*L. monocytogenes*	3.11 Log UFC/mL reduction	Stainless steel; 96 h; 37°C	SPE production by approximately 30%, inhibition of the expression of critical genes in biofilm formation such as those related to motility (*flaA, fliP, fliG, flgE, motA, motB*) and quorum sensing (*agrA, agrB, agrC*)	(Upadhyay et al., 2013)
Thymol	0.0123%	*S. aureus*	69.9%	Polystyrene; 24 h; 37°C	Not evaluated	(Nostro et al., 2007)
		S. epidermis	45.6%			
	0.05 mM	*L. monocytogenes*	3.67 Log UFC/mL reduction	Stainless steel; 96 h; 37°C	SPE production by approximately 30%, inhibition of the expression of critical genes in biofilm formation such as those related to motility (*flaA, fliP, fliG, flgE, motA, motB*) and quorum sensing (*agrA, agrB, agrC*)	(Upadhyay et al., 2013)
Trans-cinnamaldehyde	0.75 mM	*L. monocytogenes*	3.67 Log UFC/mL reduction	Stainless steel; 96 h; 37°C	Inhibition of SPE production and the expression genes related to motility (*flaA, fliP, fliG, flgE, motA, motB*) and quorum sensing (*agrA, agrB, agrC*).	(Upadhyay et al., 2013)

Table 1. (Continued)

Compound	Concentration	Bacteria	Adhesion or biofilm inhibition	Assay conditions	Mode of action	Reference
Eugenol	2.5 mM	*L. monocytogenes*	3.53 Log UFC/mL reduction	Stainless steel; 96 h; 37°C	Inhibition of SPE production and the expression genes related to motility (*flaA, fliP, fliG, flgE, motA, motB*) and quorum sensing (*agrA, agrB, agrC*).	(Upadhyay et al., 2013)
	10	*P. aeruginosa* PAO1	87%	Polystyrene; 16 h; 30°C	Competition for the binding site of *LasR*	(Sybiya Vasantha Packiavathy et al., 2012)

Adhesion and Motility

Motility mediated by flagella is critical for adhesion and biofilm formation since it allows bacteria to reach abiotic surfaces and start the attachment process (Habimana et al., 2014). For example, it has been reported that flagellum-minus and paralyzed-flagellum *L. monocytogenes* mutant strains were defective in their attachment to glass surfaces compared with the wild-type strain (Lemon et al., 2007). Similar results have been observed in *E. coli, S.* Typhimurium and *P. aeruginosa* (Almeida et al., 2012; Strehmel and Overhage, 2013). It has been proposed that the main role of flagellum during the adhesion process is to provide the energy to overcome repulsive forces among bacteria and abiotic surfaces and increase the likelihood of contact (Verstraeten et al., 2008). Thus, motility has been highlighted as a target in biofilm prevention strategies.

Plant EOs containing terpenes have been shown to inhibit bacterial adhesion to abiotic surfaces. For example, yarrow (*Achillea millefolium* L.) EO containing β-pinene, 1,8-cineole, terpinene-4-ol and caryophyllene, reduced the adhesion of *L. monocytogenes* to stainless steel (reduction of 3.34 Log CFU/cm^2) and high-density polyethylene (reduction of 2.24 Log CFU/cm^2) (Jadhav et al., 2013). Another study reports that oregano and clove EOs completely inhibited the formation of biofilms of *Sphingomonas* spp. at concentrations of 0.002 and 0.004% (w/v), respectively (Szczepanski and Lipski, 2014).

The mode of action of terpenes could be related to the down-regulation of virulence factor genes involved in the initial attachment. Upadhyay et al. (2013) found that *trans*-cinnamaldehyde, carvacrol, thymol, and eugenol reduced the adhesion of *L. monocytogenes* to stainless steel surfaces ≈ 2.5 - 3.7 Log CFU/mL and EPS production by approximately 30%. This research also evaluated the effect on transcription of genes critical for various steps involved in biofilm formation in *L. monocytogenes* and observed that these compounds down-regulated genes related to motility (*flaA, fliP, fliG, flgE, motA, motB*) and *quorum sensing* (*agrA, agrB, agrC*), stress response (*dnaK*) and transcriptional regulation (*prfA, degU, mogR*) (Upadhyay et al., 2013). Similarly, limonene, a cyclic terpene found in citrus fruits, inhibited the adhesion of *Streptococcus pyogenes, S. mutans* and *S. mitis* to polystyrene surfaces. This effect was related to the down-regulation of *covR*, *mga* and *vicR* genes, which regulate surface-associated proteins in *S. pyogenes* and *S. mutans*, respectively (Subramenium et al., 2015).

Biofilm Development

As described previously, biofilm development is characterized by the increase of EPS production that holds embedded microbial cells together, contributing to the overall architecture and resistance (Wei and Ma, 2013). In general, the main constituents in bacterial biofilms are exopolysaccharides but may also include, to a lesser extent, proteins, fractions of nucleic acids, and humic substances (Flemming and Wingender, 2010). Nevertheless, the EPS composition varies among bacterial species. For example, alginate is the exopolysaccharide that is often and mainly produced in *P. aeruginosa* biofilms (Wingender et al., 2001), whereas, in *E. coli* and *Salmonella*, cellulose is an architectural element that confers cohesion and elasticity together with the cell-encasing curli fiber network (Castelijn et al., 2012; Serra et al., 2013). In *S. aureus*, poly-N-acetylglucosamine supports cell-cell and cell-surface interactions (Harapanahalli et al., 2015). In contrast, *L. monocytogenes* biofilms have proteins as main components (Combrouse et al., 2013), including flagellin and virulence factors (internalins and ActA), that probably act as adhesins promoting cell-cell and cell-surface interactions (Franciosa et al., 2009; Guilbaud et al., 2015; Lourenço et al., 2013; Travier and Lecuit, 2014).

It has been reported that plant terpenes can affect biofilm formation by inhibiting EPS production. Nostro et al. (2007) demonstrated that oregano EO, carvacrol and thymol, at 0.5% (v/v), reduced the total biomass of biofilms formed by *S. aureus* (53.3, 71.7 and 69.9%, respectively) and *Streptococccus epidermidis* (41.4, 42.9 and 45.6, respectively). While, Tapia-Rodriguez et al. (2017) reported that carvacrol reduced the formation of *P. aeruginosa* biofilms on stainless steel surfaces in a dose-dependent manner (0.9 to 7.9 mM) cell density and biomass up to 3 Log CFU/cm^2 and 70% relative to unexposed bacteria, respectively. Also, linalool, the main constituent of orange EO, inhibited *S. aureus* biofilm formation by downregulation of genes related to the production of polysaccharide intercellular adhesion (PIA) (*icaA* and *icaB*), fibronectin-binding protein B (*fnbB*), clumping factor A and B (*clfA* and *clfB*) and elastin-binding protein (ebps) required for cell-to-cell attachment (Federman et al., 2016).

Quorum Sensing System

It has been evidenced that methyl eugenol act as an antagonist of N-3-oxo-dodecanoyl-L-homoserine lactone, the natural ligand of the LasR receptor protein in *P. aeruginosa*. Competition for the binding site interferes with intercellular communication processes affecting biofilm formation, motility, and SPE production in this bacteria (Sybiya Vasantha Packiavathy et al., 2012). Carvacrol reduced pyocyanin and violacein production in *P. aeruginosa* and *C. violaceum*, respectively; the synthesis of both pigments are regulated by quorum sensing mediated by acyl-homoserine lactones, so these findings indicate that carvacrol interferes with cell to cell communication (Tapia-Rodriguez et al., 2017). In addition, this terpene reduces the relative expression of *las*R in *P. aeruginosa* (Figure 1), causing a significant decrement of biofilm development, motility, acyl-homoserine lactones production (Tapia-Rodriguez et al., 2019). On the other hand, citral inhibited the synthesis of the autoinducer molecule AI-2 in *Salmonella enteritidis*, evidenced by using the bioluminescence reporter *Vibrio harveyi* BB170 (Zhang et al., 2014). These findings showed that the virulence of pathogenic bacteria could be affected by terpenes indicating that these bioactive compounds can modulate the behavior of bacterial communities by interfering in intercellular communication. However, future studies that deepen these interactions and their relation to the expression of genes in bacteria embedded in biofilms are required.

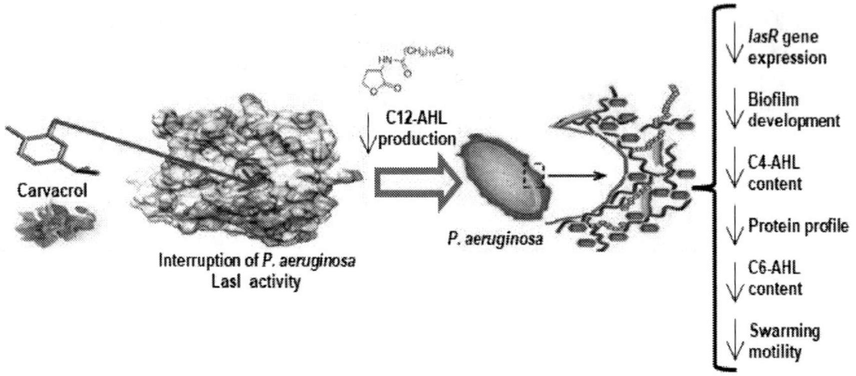

Figure 1. Carvacrol anti-*quorum sensing* potential against *P. aeruginosa.* C12-AHL: oxododecanoyl homoserine lactone, C4-AHL: butyryl homoserine lactone, C6-AHL: hexanoyl homoserine lactone.

INTERCELLULAR DAMAGES

The growing concern for the use of environmentally friendly pharmaceutical, cosmetic and agro-industrial products of natural origin, preferably vegetables, has promoted the study of the properties of plant terpenes (Ibáñez et al., 2020). Terpenes are the major compounds of EOs, composed of isoprene units with a wide variety of structural arrangements. However, 90% of the terpenes isolated from EOs are monoterpenes, for which the properties of these oils are attributed (Akthar et al., 2014).

Monoterpene phenols such as thymol and carvacrol present a benzene ring, hydroxyl, amine, or sulfhydryl substituents (Marinelli et al., 2018; Radulovic et al., 2013). The benzene ring confers lipophilic properties, electro and nucleophilic centers to monoterpenes, while the hydroxyl substituent acts as a proton donor and ionizable residue. These structural characteristics allow the interaction with various molecules and the disruption of the wall and plasma membrane in bacteria and fungi. Monoterpenes can interact directly with the membrane lipid bilayer and modulate the expression of genes coding for proteins. The integrity loss of the outer membrane implies an ionic imbalance and an increase in permeability, allowing the loss of cytoplasmic material such as ATP or nucleic acids (Marinelli et al., 2018).

Terpenes also affect the activity and expression of enzymes that intervene in the stress response of microorganisms. Among the enzymes with a considerable medical interest are drug efflux pumps (Miladi et al., 2016; Miladi et al., 2017). Antibiotic efflux pumps are one of the main resistance mechanisms among pathogens that cause nosocomial or foodborne infections. Likewise, terpenes such as β-citronellol have been reported to modulate the expression of genes involved in the adhesion process, the formation of biofilms, and the synthesis of ergosterol (Sharma et al., 2020). The ability of thymol, carvacrol, or β-citronellol to affect the membrane integrity and activity of vital enzymes of microorganisms highlights their suitability as alternative antivirulence agents.

Efflux Pump Inhibition

Drug efflux pumps are one of the main mechanisms that cause antimicrobial resistance emergence in bacteria and fungi. Efflux pumps are located in the cell membrane. They are translocators that transport drugs from the cytoplasm to the external surrounding area, causing antimicrobial concentrations to be sub-lethal in the intracellular environment and increasing the probability of generating resistance. Singh et al. (2016) reported that carvacrol or linalool could exert antimicrobial activity by disrupting the cell membrane, inhibiting the calcineurin signaling pathway, synthesizing ergosterol, or the efflux pumps' activity drugs. Yuan and Yuk (2019) evaluated the effect of sub-lethal concentrations of trans-cinnamaldehyde, thymol, and carvacrol on the virulence of *Escherichia coli* O157: H7. In this work, the authors analyzed the expression of 10 genes involved in biofilm formation, chemotaxis, flagellar development, the type III secretion system, and efflux pumps, evaluating the changes in the related virulence factors. In this investigation, they reported that the three-monoterpene phenols reduce the expression and activity of the efflux pumps without inducing important changes in the minimum inhibitory concentration of seven evaluated antibiotics.

Miladi et al. (2017) conducted a study to evaluate the antimicrobial effect of five terpenes (eugenol, carvacrol, thymol, cymene, and terpinene) and their potential role as efflux pump inhibitors of 21 food pathogenic isolates. The study reported a synergistic effect among terpenes and tetracycline, achieving a decrease of effective doses ranging from 32 to 128 μg/mL when combining these compounds. The authors suggested that these results could be enhanced by the hydrophobic nature of monoterpene phenols, which gives them a greater affinity for cell membranes. Similarly, evaluated the effect of terpenes on efflux pumps in bacteria using ethidium bromide, benzalkonium chloride, and tetracycline as indicator compounds. The increase in terpenes concentration caused an increase in the intracellular concentration of all the indicator compounds compared to the control without antimicrobials. The results reported by the authors showed that these natural compounds have a modulating effect on the activity of efflux pumps, increasing the susceptibility of pathogens to antibiotics and disinfectants.

Nakamura de Vasconcelos et al. (2018) evaluated the effect of carvacrol on efflux pumps of *Mycobacterium tuberculosis*. The impact on efflux pumps was measured by the ethidium bromide accumulation in the cell interior by adding carvacrol alone and in conjunction with rifampicin. The authors reported that the addition of carvacrol or derived compounds has an inhibitory effect on efflux pumps and generates morphological changes in bacterial cells at a concentration range of 19 to 156 μg/mL. They reported that carvacrol caused morphological changes in the cell membrane by increasing the roughness and aggregation of cells.

The effect of carvacrol and thymol on the antibiotic efflux pumps of the food pathogen, *S. aureus*, was measured by Dos Santos Barbosa et al. (2021), who report that terpenes could act as competitive inhibitors of efflux pumps. The analysis of carvacrol and thymol effects on the antibiotic norfloxacin efflux pump, NorA, showed that both compounds at 256 μg/mL decreased antibiotic efflux pump in *S. aureus*. The authors suggested that the effect of terpenes could be associated with their lipophilicity; since they can associate with the lipid bilayer and interact with the membrane proteins, including efflux pumps.

Cell Membrane Damage

Various authors have suggested that terpenes act by modifying the permeability of the bacterial cell membrane. A study by Singh et al. (2016) in 17 strains of *Candida* spp. analyzed the antifungal effect of geraniol, a monoterpene phenol identified from geranium oil. In this study, they evaluated the membrane integrity, ergosterol, and cellular susceptibility in the presence of geraniol, fluconazole, and sodium dodecyl sulfate (SDS). Singh et al. (2016) reported that geraniol increased sensitivity to membrane disrupting agents and decreased the ergosterol content. The same researchers analyzed the relationship between geraniol and the calcineurin signaling pathway, an enzyme essential for maintaining cell membrane integrity under stress conditions. They found that with the inhibition of calcineurin expression, mutant strains are more sensitive to geraniol than strains where this enzyme is overexpressed. These results suggested a relationship between the geraniol mechanism of action and the calcineurin signaling pathway.

Ansari et al. (2016) studied the relationship between calcineurin and monoterpene phenols from an approach to the *Candida* spp. genome expression changes. In this research, the authors analyzed the whole genome transcriptome changes of clinical isolates of *C. albicans* and *C.* non-*albicans* exposed to the presence of perilyl alcohol, a limonene derivative. This study revealed that perilyl alcohol has an inhibitory effect on the expression of the CNB1, a gene coding for the regulatory subunit of calcineurin. Additionally, Ansari et al. (2016) studied the isolates' cell membrane integrity by analyzing the changes in ergosterol concentration, intracellular pH, and sensitivity to cell wall disrupting agents, among other parameters. The authors report that perilyl alcohol causes a decrease of more than 50% in ergosterol concentration, an intracellular pH decrease, hypersensitivity to disrupting agents, and an increase in the cell sedimentation. Together, the research results indicate that exposure to perilyl alcohol generated cell wall and membrane damage in the clinical isolates of *Candida* spp.

Another suggested mechanism through which terpenes affect the integrity of cell membranes is by inhibiting the synthesis of ergosterol, the main fungal sterol, located mainly in the cytoplasmic layer of the fungal cell membrane (Solanko et al., 2018). Pereira et al. (2015) evaluated the effect of geraniol and β-citronellol on the cell wall and membrane and the biosynthesis of ergosterol *Trichophyton rubrum*. These authors reported that both compounds caused intracellular material loss and decreased intracellular ergosterol concentration, suggesting inhibition of its biosynthesis. However, a recent study by Sharma et al. (2020) evaluated the β-citronellol antifungal action on modified strains of *C. albicans* through susceptibility tests and genes involved in the ergosterol biosynthesis. However, the new results reported by Sharma et al. (2020) contradict the theory proposed by Pereira et al. (2015). Sharma et al. (2020) found that β-citronellol increases cellular sensitivity to membrane disrupting agents, such as fluconazole or SDS. Despite that, the evaluation of ergosterol biosynthesis genes (*erg1, erg3,* and *erg11*) showed that β-citronellol promoted the overexpression of *erg1, erg3*, and inhibited *erg11*. Sharma et al. (2020) suggested that oxygen variations modified the ERG genes expression; therefore, the anaerobic environment caused by the β-citronellol addition could limit the gene expression results.

Authors have suggested that monoterpenes also act directly at the cell membrane since terpenes are hydrophobic molecules with an affinity for the lipid bilayer. For example, the carvacrol octanol-water partition coefficient is 3.64, implying a high affinity for animal fat tissues (Sharifi-Rad et al., 2018). Moreover, geraniol, limonene, linalool, or carvacrol have a hydroxyl group that could interact with the carbon chains of cell membrane lipids. Silva et al. (2017) studied the effect of linalool on *Microsporum* spp. clinical isolates measured the loss of intracellular content and the extracellular potassium concentration changes. The authors reported that a linalool concentration of 128 μg/mL caused a significant intracellular material loss and an efflux of potassium ions. Those results indicated that linalool caused destabilization of the cell membrane in *Microsporum* spp. and it could be used as an antimicrobial agent to treat fungi infections. All of the above evidence shows that even when it is known that terpenes disrupt the cell membrane, affecting ergosterol levels and inhibiting calcineurin signaling pathway, still is necessary to carry out future research to deepen their mechanism of action.

Interactions with Bacterial Cell Proteins and Nucleic Acid

Terpenes cause bacterial death by inhibiting proteins' synthesis and functionality. A study conducted by Guo et al. (2021b) demonstrated that linalool seriously affected the metabolic pathways related to protein synthesis in the bacterium *Shewanella putrefaciens*.

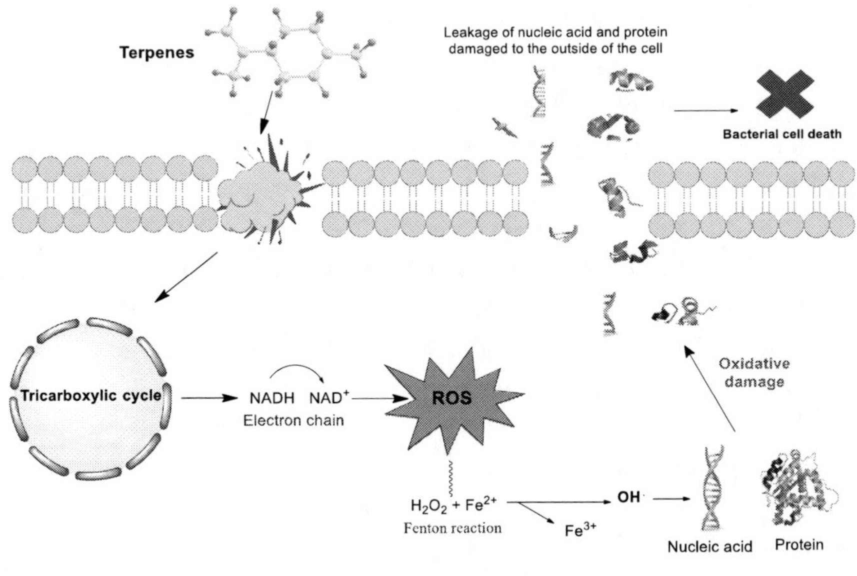

Figure 2. Proposed action mechanism for terpenes damage to nucleic acids and proteins.

Pantothenate and CoA are key cofactors of bacterial cellular metabolism (Hussein et al., 2020). Pantothenate is the precursor to CoA and the basis of acyl transporter protein (ACP) repair, which is involved in the metabolic pathway of proteins in organisms (Shu et al., 2020). Terpenes can cause bacterial death by damaging intracellular proteins and nucleic acids (Figure 2).

Bacterial Cell Proteins

Guo et al. (2021b) found that the treatment of linalool on *S. putrefaciens* decreased the pantothenate content, and the precursors related to the biosynthesis of pantothenate and CoA were significantly modified (Figure 3). In addition, there were significant changes in the biosynthesis of amino acid-RNAt, which is part of the protein translation pathway. Interference in the amino acid-RNAt synthesis pathway affects protein synthesis, cell proliferation, and signal transduction (Gan et al., 2020).

In another study in which linalool was used as an antibacterial agent against *Pseudomonas fluorescens,* this treatment inhibited cellular respiration and energy synthesis by decreasing enzyme activity, including succinate dehydrogenase, malate dehydrogenase, pyruvate kinase, and ATPase (Qiu et al., 2011). Therefore, linalool is proposed as a food antiseptic in food systems. Some terpenes such as menthol decreased the expression of extracellular proteins associated with the virulence of *S. aureus.* It has been shown that applying menthol on methicillin-sensitive, and methicillin-resistant *S. aureus* can substantially inhibit α-toxin, enterotoxins SEA, SEB, and TSST-1 expression in methicillin-sensitive *Staphylococcus aureus* (MSSA) and methicillin-resistant *Staphylococcus aureus* (MRSA) (Qiu et al., 2011). This result suggests that menthol has the potential as an anti-virulence agent.

When *E. coli* O157:H7 was exposed to carvacrol (1 mM) for 16 h, it inhibited flagellin synthesis very significantly ($P < 0.001$), causing non-flagellated and non-mobile cells. The use of carvacrol against *E. coli* O157:H7 may be advantageous because a-flagellated cells have a lower ability to adhere to epithelial cells and are less invasive than flagellated cells (Burt et al., 2007). For this reason, it is possible to inhibit the expression of certain virulence factors and the pathogenicity of bacteria.

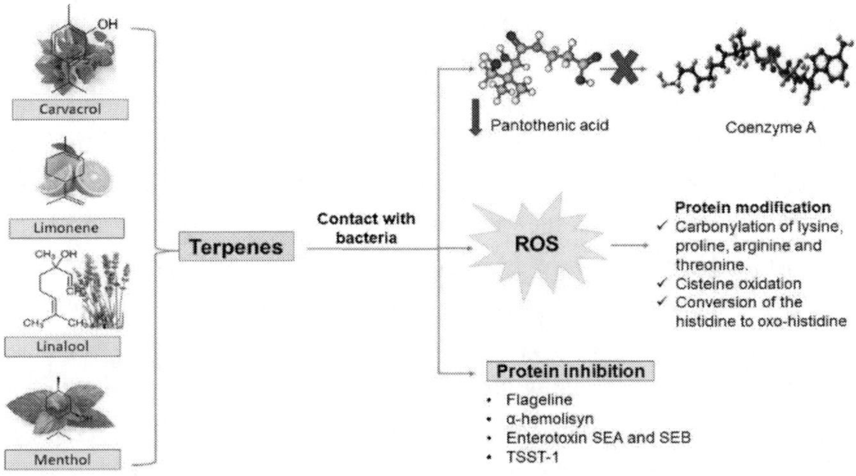

Figure 3. Mechanism of action de terpenes to intervene in the synthesis and inhibition of bacterial proteins.

Nucleic Acids

Chueca et al. (2014a) suggested that the inactivation of *E. coli* in the presence of carvacrol (300 μL/L) and citral (100 μL/L) was mediated by reactive oxygen species (ROS). Among ROS, hydroxyl radicals are the most reactive species that can cause oxidative DNA damage, causing lethal double-stranded mutations (Foti et al., 2012). Hydroxyl radicals could be generated through the reaction of hydrogen peroxide with other compounds or other ROS involved in DNA damage. Resulting in the formation of toxic byproducts, such as different aldehydes derived from ROS, causing DNA damage (Esterbauer et al., 1991). Citral and carvacrol promoted the production of intracellular ROS in bacteria under aerobic conditions, causing direct damage to cellular components and bacterial death (Chueca et al., 2014b).

Within ROS is the hydroxyl ion (OH^-), hydrogen peroxide (H_2O_2), peroxide (O_2^{2-}), singlet oxygen ($1O_2$), hydroxyl radical ($OH\cdot$), and superoxide anion (O_2^-); these generally affect lipids, proteins and nucleic acids (Memar et al., 2018; Pizzino et al., 2017). At the bacterial level, such a reaction will cause damage to the membrane and eventually kill the cells. The interaction between ROS and proteins generally destabilizes and inactivates a particular protein due to the covalent modifications they cause. Such changes include carbonylation of lysine, proline, arginine, and threonine, oxidation of cysteine, and conversion of histidine to oxo-histidine, which affects the protein function (Ezraty et al., 2017). In addition, another ROS target is guanine, which causes oxidation and DNA breakage. All these events lead to the production of non-functional proteins and loss of cellular viability.

Another monoterpene with antimicrobial activity is (+)-limonene, extracted from citrus EOs. Using this monoterpene at a concentration of 2,000 μL/L causes inactivation of *E. coli* due to the tricarboxylic acid and Fenton cycle mediated by the formation of hydroxyl radicals that causes oxidative damage to bacterial DNA, affecting nucleic acids replication (Chueca et al., 2014a). Subsequently, Zuniga et al. (2006) related the DNA damage observed in bacteria treated with terpenes and catechins with reactive oxygen potential.

Several investigations have shown that thymol can cause damage to the cell membrane and cause the leakage of intracellular components, nucleic acids. This effect was demonstrated by a study conducted by Cai et al. (2019) evaluating the effect of thymol (0.25-0.5 mg/mL) against vegetative cells and spores of *Alicyclobacillus acidoterrestris* for 2, 6, and 24 h. In the presence of thymol (0.25 mg/mL), nucleic acid leakage from *A. acidoterrestris* increased 1.7, 1.6, and 1.3 times concerning control, for 2.6 and 24 h, respectively. In addition, with the application of thymol for 2 h, a protein leakage from the vegetative bacterial cells of 1.13 mg/mL was observed. These results suggested that thymol damaged *A. acidoterrestris,* which caused leakage of nucleic acids and proteins. The hydroxyl group of thymol can interact with the bacterial membrane and cause its alteration, resulting in the leakage of cellular components

(Hyldgaard et al., 2012). For these reasons, thymol treatments can be used against *A. acidoterrestris* to control its presence in the food industry. When investigating the mechanism of carvacrol against *S. aureus*, the results showed that this compound bound to the lower groove of DNA and that hydrogen bonds played an important role in the interaction of carvacrol with the DNA of *S. aureus* (Wang et al., 2016). Some of the actions of carvacrol's antibacterial activity may be related to its DNA binding ability.

BACTERIOSTATIC TERPENES MODE OF ACTION

Previously, the antibacterial characteristics of terpenes have been described. A bacteriostatic compound stops or inhibits cellular growth; on the other hand, the bactericidal compound kills the cell (Zengin and Baysal, 2014); other differences are shown in table 2. The bacteriostatic or bactericidal effect depends on three characteristics: terpene, medium composition, and environmental factors (Munekata et al., 2020). This section described the mode of action of bacteriostatic terpenes.

Table 2. Relevant differences between bactericidal and bacteriostatic terpenes

Bactericidal terpenes	Bacteriostatic terpenes
Kill the bacteria	Do not kill the bacteria
Irreversible damage	Reversible damage
Induce membrane disruption	Do not induce membrane disruption
Leakage of ions	Inhibits the microbial growth
Affect intracellular components	Do not affect intracellular components
Related with Minimum Bactericide Concentration	Related with Minimum Inhibitory Concentration

A terpene can present both a bacteriostatic effect at low concentrations and a bactericidal effect at high concentrations (Mourey and Canillac, 2002). In this sense, terpenes of conifer essential oil showed that concentrations are the key for one or other effects, where high concentrations showed a bactericidal effect and low concentrations were bacteriostatic.

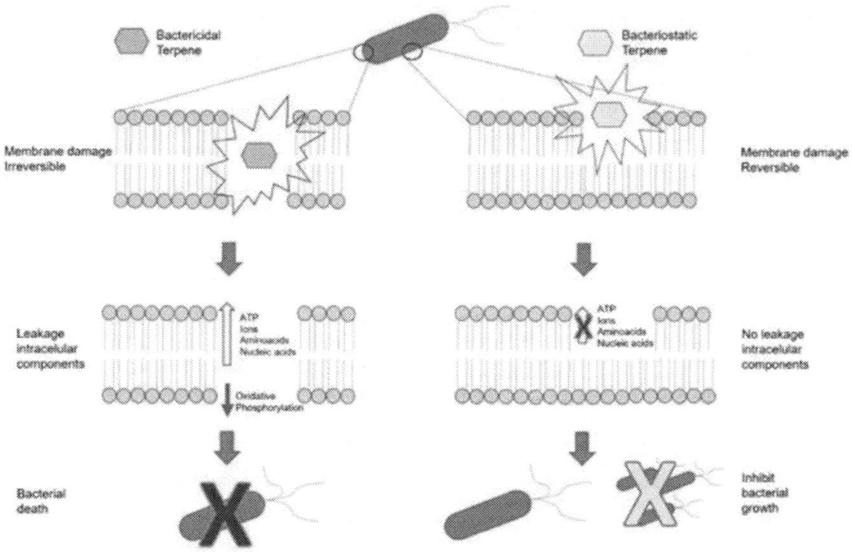

Figure 4. Bactericidal terpene and bacteriostatic terpene: Principal mode of action.

Bacteriostatic terpenes can be ineffective as bactericidal agents, but they can affect the expression of virulence factors. Terpenes with bacteriostatic activity do not penetrate the outer membrane in Gram-negative bacteria and do not affect the intracellular ATP concentration causing a bacteriostatic effect as a principal mode of action (Gallucci et al., 2009; Guimarães et al., 2019). In Gram-positive bacteria, this effect is similar, Togashi et al. (2008) found that terpenes as geraniol and farnesol inhibited bacterial growth, but disruption in cell membrane was not observed (Figure 4). Furthermore, other terpenes have another mode of action; these access the periplasm and inhibited the transmembrane ATPase activity, causing ATP depletion and impeding bacterial growth but not cellular viability (Hyldgaard et al., 2012). Bacteriostatic terpenes have a mode of action that allows the cell to remain alive but unable to reproduce since their energy reserves and vital components are being used to repair the damage. A bacterium treated with terpenes or other antibacterial compounds requires time to repair the damages; these actions cause the inhibited growth, and sometimes it is reflected by an elongation of the lag phase (Mazzarrino et al., 2015).

Conclusion

Terpenes compounds can be used as an anti-virulence alternative, but more studies are still needed to understand their mechanism of action. This knowledge will expand and improve the use of these compounds as natural anti-virulence agents. In this context, research of terpenes is needed in bacterial metabolic changes at the molecular level to develop food-related bacteriostatic agents and establish how to direct terpenes to a specific site to decrease or inhibit virulence genes expression.

References

Akthar, M. S., Degaga, B., and Azam, T. 2014. Antimicrobial activity of essential oils extracted from medicinal plants against the pathogenic microorganisms: A review. *Journal Issues ISSN*. 2350:1588.

AlAbbas, F. M., Spear, J. R., Kakpovbia, A., Balhareth, N. M., Olson, D. L., and Mishra, B. 2012. Bacterial attachment to metal substrate and its effects on microbiologically-influenced corrosion in transporting hydrocarbon pipelines. *Journal of Pipeline Engineering*. 11:63.

Almeida, C., Nóbrega, F., Kluskens, L., Azevedo, N., Keevil, C., and Vieira, M. 2012. *The role of flagellum and flagellum-based motility on Salmonella Enteritidis and Escherichia coli biofilm formation.*

Alvarez, M. V., Ortega-Ramirez, L. A., Gutierrez-Pacheco, M. M., Bernal-Mercado, A. T., Rodriguez-Garcia, I., Gonzalez-Aguilar, G. A., Ponce, A., Moreira, M. d. R., Roura, S. I., and Ayala-Zavala, J. F. 2014. Oregano essential oil-pectin edible films as anti-quorum sensing and food antimicrobial agents. *Frontiers in Microbiology*. 5:699.

Ansari, M. A., Fatima, Z., and Hameed, S. 2016. Anticandidal effect and mechanisms of monoterpenoid, perillyl alcohol against *Candida albicans*. *PloS one*. 11:e0162465.

Borges, A., Abreu, A., Malheiro, J., Saavedra, M. J., and Simõe, M. 2013. Biofilm prevention and control by dietary phytochemicals. *CECAV-Centro de Ciência Animal e Veterinária*. 3:2-41.

Burt, S. A., van der Zee, R., Koets, A. P., de Graaff, A. M., van Knapen, F., Gaastra, W., Haagsman, H. P., and Veldhuizen, E. J. 2007. Carvacrol induces heat shock protein 60 and inhibits synthesis of flagellin in Escherichia coli O157: H7. *Applied and environmental microbiology*. 73:4484-4490.

Cai, R., Zhang, M., Cui, L., Yuan, Y., Yang, Y., Wang, Z., and Yue, T. 2019. Antibacterial activity and mechanism of thymol against Alicyclobacillus acidoterrestris vegetative cells and spores. *LWT*. 105:377-384.

Castelijn, G. A. A., van der Veen, S., Zwietering, M. H., Moezelaar, R., and Abee, T. 2012. Diversity in biofilm formation and production of curli fimbriae and cellulose of *Salmonella* Typhimurium strains of different origin in high and low nutrient medium. *Biofouling*. 28:51-63.

Combrouse, T., Sadovskaya, I., Faille, C., Kol, O., Guérardel, Y., and Midelet-Bourdin, G. 2013. Quantification of the extracellular matrix of the *Listeria monocytogenes* biofilms of different phylogenic lineages with optimization of culture conditions. *Journal of Applied Microbiology*. 114:1120-1131.

Chueca, B., Pagan, R., and Garcia-Gonzalo, D. 2014a. Differential mechanism of Escherichia coli inactivation by (+)-limonene as a function of cell physiological state and drug's concentration. *PLoS One*. 9:e94072.

Chueca, B., Pagán, R., and García-Gonzalo, D. 2014b. Oxygenated monoterpenes citral and carvacrol cause oxidative damage in Escherichia coli without the involvement of tricarboxylic acid cycle and Fenton reaction. *International journal of food microbiology*. 189:126-131.

Donlan, R. M. 2002. Biofilms: microbial life on surfaces. *Emerging Infectious Diseases*. 8:881-890.

Dos Santos Barbosa, C. R., Scherf, J. R., de Freitas, T. S., de Menezes, I. R. A., Pereira, R. L. S., dos Santos, J. F. S., de Jesus, S. S. P., Lopes, T. P., de Sousa Silveira, Z., and de Morais Oliveira-Tintino, C. D. 2021. Effect of Carvacrol and Thymol on NorA efflux pump inhibition in multidrug-resistant (MDR) *Staphylococcus aureus* strains. *Journal of Bioenergetics and Biomembranes*:1-10.

Esterbauer, H., Schaur, R. J., and Zollner, H. 1991. Chemistry and biochemistry of 4-hydroxynonenal, malonaldehyde and related aldehydes. *Free radical Biology and medicine*. 11:81-128.

Ezraty, B., Gennaris, A., Barras, F., and Collet, J. F. 2017. Oxidative stress, protein damage and repair in bacteria. *Nature Reviews Microbiology*. 15:385-396.

Federman, C., Ma, C., and Biswas, D. 2016. Major components of orange oil inhibit *Staphylococcus aureus* growth and biofilm formation, and alter its virulence factors. *Journal of Medical Microbiology*. 65:688-695.

Flemming, H. C., and Wingender, J. 2010. The biofilm matrix. *Nature Reviews Microbiology*. 8:623-633.

Floyd, K. A., Eberly, A. R., and Hadjifrangiskou, M. 2017. Adhesion of bacteria to surfaces and biofilm formation on medical devices. In *Biofilms and Implantable Medical Devices*. Y. Deng and W. Lv, editors. Woodhead Publishing. 47-95.

Foti, J. J., Devadoss, B., Winkler, J. A., Collins, J. J., and Walker, G. C. 2012. Oxidation of the guanine nucleotide pool underlies cell death by bactericidal antibiotics. *Science*. 336:315-319.

Franciosa, G., Maugliani, A., Scalfaro, C., Floridi, F., and Aureli, P. 2009. Expression of internalin a and biofilm formation among *Listeria monocytogenes* clinical isolates. *International Journal of Immunopathology and Pharmacology*. 22:183-193.

Gallucci, M. N., Oliva, M., Casero, C., Dambolena, J., Luna, A., Zygadlo, J., and Demo, M. 2009. Antimicrobial combined action of terpenes against the foodborne microorganisms *Escherichia coli,*

Staphylococcus aureus and *Bacillus cereus*. *Flavour and Fragrance Journal*. 24:348-354.

Gan, C., Huang, X., Wu, Y., Zhan, J., Zhang, X., Liu, Q., and Huang, Y. 2020. Untargeted metabolomics study and pro-apoptotic properties of B-norcholesteryl benzimidazole compounds in ovarian cancer SKOV3 cells. *The Journal of Steroid Biochemistry and Molecular Biology*. 202:105709.

Guilbaud, M., Piveteau, P., Desvaux, M., Brisse, S., and Briandet, R. 2015. Exploring the diversity of *Listeria monocytogenes* biofilm architecture by high-throughput confocal laser scanning microscopy and the predominance of the honeycomb-like morphotype. *Applied and Environmental Microbiology*. 81:1813-1819.

Guimarães, A. C., Meireles, L. M., Lemos, M. F., Guimarães, M. C. C., Endringer, D. C., Fronza, M., and Scherer, R. 2019. Antibacterial activity of terpenes and terpenoids present in essential oils. *Molecules*. 24:2471.

Guo, F., Chen, Q., Liang, Q., Zhang, M., Chen, W., Chen, H., Yun, Y., Zhong, Q., and Chen, W. 2021a. Antimicrobial Activity and Proposed Action Mechanism of Linalool against *Pseudomonas fluorescens. Frontiers in Microbiology*. 12:49.

Guo, F., Liang, Q., Zhang, M., Chen, W., Chen, H., Yun, Y., Zhong, Q., and Chen, W. 2021b. Antibacterial activity and mechanism of linalool against *Shewanella putrefaciens*. *Molecules*. 26:245.

Gutierrez-Pacheco, M. M., Bernal-Mercado, A. T., Vazquez-Armenta, F. J., González-Aguilar, G., Lizardi-Mendoza, J., Madera-Santana, T., Nazzaro, F., Ayala-Zavala, J. J. P., and Pathology, M. P. 2019. Quorum sensing interruption as a tool to control virulence of plant pathogenic bacteria. 106:281-291.

Habimana, O., Semião, A., and Casey, E. 2014. The role of cell-surface interactions in bacterial initial adhesion and consequent biofilm formation on nanofiltration/reverse osmosis membranes. *Journal of Membrane Science*. 454:82-96.

Harapanahalli, A. K., Chen, Y., Li, J., Busscher, H. J., and van der Mei, H. C. 2015. Influence of Adhesion Force on icaA and cidA Gene

Expression and Production of Matrix Components in *Staphylococcus aureus* Biofilms. *Applied and Environmental Microbiology*. 81:3369-3378.

Hobley, L., Harkins, C., MacPhee, C. E., and Stanley-Wall, N. R. 2015. Giving structure to the biofilm matrix: an overview of individual strategies and emerging common themes. *FEMS Microbiology Reviews*. 39:649-669.

Hussein, M., Schneider-Futschik, E. K., Paulin, O. K., Allobawi, R., Crawford, S., Zhou, Q. T., Hanif, A., Baker, M., Zhu, Y., and Li, J. 2020. Effective strategy targeting polymyxin-resistant Gram-negative pathogens: Polymyxin B in combination with the selective serotonin reuptake inhibitor sertraline. *ACS Infectious Diseases*. 6:1436-1450.

Hyldgaard, M., Mygind, T., and Meyer, R. L. 2012. Essential oils in food preservation: mode of action, synergies, and interactions with food matrix components. *Frontiers in Microbiology*. 3:12.

Ibáñez, M. D., López-Gresa, M. P., Lisón, P., Rodrigo, I., Bellés, J. M., González-Mas, M. C., and Blázquez, M. A. 2020. Essential oils as natural antimicrobial and antioxidant products in the agrifood industry. *Nereis. Interdisciplinary Ibero-American Journal of Methods, Modelling and Simulation*: 55-69.

Jadhav, S., Shah, R., Bhave, M., and Palombo, E. A. 2013. Inhibitory activity of yarrow essential oil on *Listeria* planktonic cells and biofilms. *Food Control*. 29:125-130.

Lemon, K. P., Higgins, D. E., and Kolter, R. 2007. Flagellar motility is critical for *Listeria monocytogenes* biofilm formation. *Journal of Bacteriology*. 189:4418-4424.

Lourenço, A., de Las Heras, A., Scortti, M., Vazquez-Boland, J., Frank, J. F., and Brito, L. 2013. Comparison of *Listeria monocytogenes* exoproteomes from biofilm and planktonic state: lmo2504, a protein associated with biofilms. *Applied and Environmental Microbiology*. 79:6075-6082.

Luna-Solorza, J. M., Vazquez-Armenta, F. J., Bernal-Mercado, A. T., Gutierrez-Pacheco, M. M., Nazzaro, F., and Ayala-Zavala, J. F. 2020. Phytochemical compounds targeting the quorum sensing system as a tool to reduce the virulence factors of food pathogenic bacteria. *Trends in Quorum Sensing and Quorum Quenching*: 257-276.

Marinelli, L., Di Stefano, A., and Cacciatore, I. 2018. Carvacrol and its derivatives as antibacterial agents. *Phytochemistry Reviews*. 17:903-921.

Mazzarrino, G., Paparella, A., Chaves-López, C., Faberi, A., Sergi, M., Sigismondi, C., Compagnone, D., and Serio, A. 2015. *Salmonella enterica* and *Listeria monocytogenes* inactivation dynamics after treatment with selected essential oils. *Food Control*. 50:794-803.

Memar, M. Y., Ghotaslou, R., Samiei, M., and Adibkia, K. 2018. Antimicrobial use of reactive oxygen therapy: current insights. *Infection and drug resistance*. 11:567.

Miladi, H., Zmantar, T., Chaabouni, Y., Fedhila, K., Bakhrouf, A., Mahdouani, K., and Chaieb, K. 2016. Antibacterial and efflux pump inhibitors of thymol and carvacrol against foodborne pathogens. *Microbial Pathogenesis*. 99:95-100.

Miladi, H., Zmantar, T., Kouidhi, B., Al Qurashi, Y. M. A., Bakhrouf, A., Chaabouni, Y., Mahdouani, K., and Chaieb, K. 2017. Synergistic effect of eugenol, carvacrol, thymol, p-cymene and γ-terpinene on inhibition of drug resistance and biofilm formation of oral bacteria. *Microbial Pathogenesis*. 112:156-163.

Mourey, A., and Canillac, N. 2002. Anti-*Listeria monocytogenes* activity of essential oils components of conifers. *Food control*. 13:289-292.

Munekata, P. E., Pateiro, M., Rodríguez-Lázaro, D., Domínguez, R., Zhong, J., and Lorenzo, J. M. 2020. The role of essential oils against pathogenic *Escherichia coli* in food products. *Microorganisms*. 8:924.

Nakamura de Vasconcelos, S. S., Caleffi-Ferracioli, K. R., Hegeto, L. A., Baldin, V. P., Nakamura, C. V., Stefanello, T. F., Freitas Gauze, G. d., Yamazaki, D. A., Scodro, R. B., and Siqueira, V. L. 2018. Carvacrol activity and morphological changes in *Mycobacterium tuberculosis*. *Future Microbiology*. 13:877-888.

Nazzaro, F., Fratianni, F., and Coppola, R. 2013. Quorum sensing and phytochemicals. *International Journal of Molecular Sciences*. 14:12607.

Nostro, A., Cellini, L., Zimbalatti, V., Blanco, A. R., Marino, A., Pizzimenti, F., Giulio, M. D., and Bisignano, G. 2012. Enhanced activity of carvacrol against biofilm of *Staphylococcus aureus* and *Staphylococcus epidermidis* in an acidic environment. *Apmis*. 120:967-973.

Nostro, A., Roccaro, A. S., Bisignano, G., Marino, A., Cannatelli, M. A., Pizzimenti, F. C., Cioni, P. L., Procopio, F., and Blanco, A. R. 2007. Effects of oregano, carvacrol and thymol on *Staphylococcus aureus* and *Staphylococcus epidermidis* biofilms. *Journal of Medical Microbiology*. 56:519-523.

Pereira, F. d. O., Mendes, J. M., Lima, I. O., Mota, K. S. d. L., Oliveira, W. A. d., and Lima, E. d. O. 2015. Antifungal activity of geraniol and citronellol, two monoterpenes alcohols, against *Trichophyton rubrum* involves inhibition of ergosterol biosynthesis. *Pharmaceutical Biology*. 53:228-234.

Pizzino, G., Irrera, N., Cucinotta, M., Pallio, G., Mannino, F., Arcoraci, V., Squadrito, F., Altavilla, D., and Bitto, A. 2017. Oxidative stress: harms and benefits for human health. *Oxidative medicine and cellular longevity*. 2017.

Qiu, J., Luo, M., Dong, J., Wang, J., Li, H., Wang, X., Deng, Y., Feng, H., and Deng, X. 2011. Menthol diminishes *Staphylococcus aureus* virulence-associated extracellular proteins expression. *Applied microbiology and biotechnology*. 90:705-712.

Radulovic, N., Blagojevic, P., Stojanovic-Radic, Z., and Stojanovic, N. 2013. Antimicrobial plant metabolites: structural diversity and mechanism of action. *Current Medicinal Chemistry*. 20:932-952.

Serra, D. O., Richter, A. M., and Hengge, R. 2013. Cellulose as an architectural element in spatially structured *Escherichia coli* biofilms. *Journal of Bacteriology*. 195:5540-5554.

Sharifi-Rad, M., Varoni, E. M., Iriti, M., Martorell, M., Setzer, W. N., del Mar Contreras, M., Salehi, B., Soltani-Nejad, A., Rajabi, S., and Tajbakhsh, M. 2018. Carvacrol and human health: A comprehensive review. *Phytotherapy Research*. 32:1675-1687.

Sharma, Y., Rastogi, S. K., Perwez, A., Rizvi, M. A., and Manzoor, N. 2020. β-citronellol alters cell surface properties of *Candida albicans* to influence pathogenicity related traits. *Medical Mycology*. 58:93-106.

Shu, H., Zhang, W., Yun, Y., Chen, W., Zhong, Q., Hu, Y., Chen, H., and Chen, W. 2020. Metabolomics study on revealing the inhibition and metabolic dysregulation in *Pseudomonas fluorescens* induced by 3-carene. *Food Chemistry*. 329:127220.

Silva, K. V., Lima, M. I., Cardoso, G. N., Santos, A. S., Silva, G. S., and Pereira, F. O. 2017. Inibitory effects of linalool on fungal pathogenicity of clinical isolates of *Microsporum canis* and *Microsporum gypseum*. *Mycoses*. 60:387-393.

Singh, S., Fatima, Z., and Hameed, S. 2016. Insights into the mode of action of anticandidal herbal monoterpenoid geraniol reveal disruption of multiple MDR mechanisms and virulence attributes in *Candida albicans*. *Archives of Microbiology*. 198:459-472.

Solanko, L. M., Sullivan, D. P., Sere, Y. Y., Szomek, M., Lunding, A., Solanko, K. A., Pizovic, A., Stanchev, L. D., Pomorski, T. G., and Menon, A. K. 2018. Ergosterol is mainly located in the cytoplasmic leaflet of the yeast plasma membrane. *Traffic*. 19:198-214.

Srey, S., Jahid, I. K., and Ha, S. D. 2013. Biofilm formation in food industries: A food safety concern. *Food Control*. 31:572-585.

Strehmel, J., and Overhage, J. 2013. The sensor kinase PA4398 regulates swarming motility and biofilm formation in *Pseudomonas aeruginosa* PA14. *YIN-Day 2013 Posters*: 3.

Subramenium, G. A., Vijayakumar, K., and Pandian, S. K. 2015. Limonene inhibits streptococcal biofilm formation by targeting surface-associated virulence factors. *Journal of Medical Microbiology*. 64:879-890.

Sybiya Vasantha Packiavathy, I. A., Agilandeswari, P., Musthafa, K. S., Karutha Pandian, S., and Veera Ravi, A. 2012. Antibiofilm and quorum sensing inhibitory potential of *Cuminum cyminum* and its secondary metabolite methyl eugenol against Gram negative bacterial pathogens. *Food Research International*. 45:85-92.

Szczepanski, S., and Lipski, A. 2014. Essential oils show specific inhibiting effects on bacterial biofilm formation. *Food Control*. 36:224-229.

Tapia-Rodriguez, M. R., Bernal-Mercado, A. T., Gutierrez-Pacheco, M. M., Vazquez-Armenta, F. J., Hernandez-Mendoza, A., Gonzalez-Aguilar, G. A., Martinez-Tellez, M. A., Nazzaro, F., and Ayala-Zavala, J. F. 2019. Virulence of *Pseudomonas aeruginosa* exposed to carvacrol: Alterations of the Quorum sensing at enzymatic and gene levels. *Journal of Cell Communication and Signaling*. 13:531-537.

Tapia-Rodriguez, M. R., Hernandez-Mendoza, A., Gonzalez-Aguilar, G. A., Martinez-Tellez, M. A., Martins, C. M., and Ayala-Zavala, J. F. 2017. Carvacrol as potential quorum sensing inhibitor of *Pseudomonas aeruginosa* and biofilm production on stainless steel surfaces. *Food Control*. 75:255-261.

Togashi, N., Inoue, Y., Hamashima, H., and Takano, A. 2008. Effects of two terpene alcohols on the antibacterial activity and the mode of action of farnesol against *Staphylococcus aureus*. *Molecules*. 13:3069-3076.

Travier, L., and Lecuit, M. 2014. *Listeria monocytogenes* ActA: a new function for a 'classic' virulence factor. *Current Opinion in Microbiology*. 17:53-60.

Upadhyay, A., Upadhyaya, I., Kollanoor-Johny, A., and Venkitanarayanan, K. 2013. Antibiofilm effect of plant derived antimicrobials on *Listeria monocytogenes*. *Food Microbiology*. 36:79-89.

Veerachamy, S., Yarlagadda, T., Manivasagam, G., and Yarlagadda, P. K. 2014. Bacterial adherence and biofilm formation on medical implants: A review. *Proceedings of the Institution of Mechanical Engineers, Part H: Journal of Engineering in Medicine*. 228:1083-1099.

Verstraeten, N., Braeken, K., Debkumari, B., Fauvart, M., Fransaer, J., Vermant, J., and Michiels, J. 2008. Living on a surface: swarming and biofilm formation. *Trends in Microbiology*. 16:496-506.

Wang, L. H., Wang, M. S., Zeng, X. A., Zhang, Z. H., Gong, D. M., and Huang, Y. B. 2016. Membrane destruction and DNA binding of Staphylococcus aureus cells induced by carvacrol and its combined effect with a pulsed electric field. *Journal of agricultural and food chemistry*. 64:6355-6363.

Wei, Q., and Ma, L. 2013. Biofilm matrix and its regulation in *Pseudomonas aeruginosa*. *International Journal of Molecular Sciences*. 14:20983.

Wingender, J., Strathmann, M., Rode, A., Leis, A., and Flemming, H. C. 2001. Isolation and biochemical characterization of extracellular polymeric substances from *Pseudomonas aeruginosa*. In *Methods in Enzymology*. Vol. Volume 336. J. D. Ron, editor. Academic Press. 302-314.

Wittry, B., and Nicholas, D. 2020. Modernizing the Foodborne Outbreak Contributing Factors: The Key to Prevention. *Journal of Environmental Health*. 83:42-45.

Yuan, W., and Yuk, H. G. 2019. Effects of sublethal thymol, carvacrol, and trans-cinnamaldehyde adaptation on virulence properties of *Escherichia coli* O157: H7. *Applied and Environmental Microbiology*. 85:e00271-00219.

Zengin, H., and Baysal, A. H. 2014. Antibacterial and antioxidant activity of essential oil terpenes against pathogenic and spoilage-forming bacteria and cell structure-activity relationships evaluated by SEM microscopy. *Molecules*. 19:17773-17798.

Zeraik, A. E., and Nitschke, M. 2012. Influence of growth media and temperature on bacterial adhesion to polystyrene surfaces. *Brazilian Archives of Biology and Technology*. 55:569-576.

Zhang, H., Zhou, W., Zhang, W., Yang, A., Liu, Y., Jiang, Y., Huang, S., and Su, J. 2014. Inhibitory effects of citral, cinnamaldehyde, and tea polyphenols on mixed biofilm formation by foodborne *Staphylococcus aureus* and *Salmonella enteritidis*. *Journal f Food Protection*. 77:927-933.

Zhu, Y., Zhang, Y., Ren, H. Q., Geng, J. J., Xu, K., Huang, H., and Ding, L. L. 2015. Physicochemical characteristics and microbial community evolution of biofilms during the start-up period in a moving bed biofilm reactor. *Bioresource Technology*. 180:345-351.

Zuniga, M. A., Dai, J., Wehunt, M. P., and Zhou, Q. 2006. DNA oxidative damage by terpene catechols as analogues of natural terpene quinone methide precursors in the presence of Cu (II) and/or NADH. *Chemical research in toxicology*. 19:828-836.

In: Pathogenic Bacteria
Editor: Keith D. Watts
ISBN: 978-1-68507-422-7

Chapter 4

ANTIPATHOGENIC POTENTIAL OF *FERULA ASAFOETIDA* ESSENTIAL OIL

Sanjay Joshi[1], Pinakin Khambhala[1], Mayank Shah[1], Shailja Varma[1], Gemini Gajera[2], Sriram Seshadri[2] and Vijay Kothari[2,*]

[1]National Foods, Vadodara, India
[2]Institute of Science, Nirma University, Ahmedabad, India

ABSTRACT

Asafoetida is a plant being used since hundreds of years for edible as well as medicinal purposes. We undertook to investigate anti-pathogenic potential of essential oil from this plant obtained through microwave-assisted extraction method. This oil was found to be able to inhibit growth and quorum sensing-regulated pigment formation of pathogenic bacteria like *Pseudomonas aeruginosa* and *Chromobacterium violaceum*. This oil (from Kabuli variety) was able to reduce virulence of *P. aeruginosa* towards the model host *Caenorhabditis elegans*. Repeated exposure of *P. aeruginosa* to

* Corresponding Author's E-mail: vijay.kothari@nirmauni.ac.in.

asafoetida oil was not found to induce any resistance in this bacterium against this oil's anti-infective activity. *Iran, Shiro,* and *Uzbeki* varieties of asafoetida were able to inhibit growth of *Shigella flexneri, Staphylococcus hominis,* and *Vibrio cholerae*. Various phenotypic traits (e.g., biofilm formation, antibiotic susceptibility, and catalase activity) of some of these pathogenic bacteria were also modulated under influence of asafoetida oil. Though our hitherto experiments indicate asafoetida oil to possess anti-pathogenic activity, in most cases, such activity is exhibited at relatively higher concentrations, and hence it seems necessary to isolate the bioactive fractions/compounds from this oil for their individual assessment with respect to potential antibacterial activity.

Keywords: *Ferula asafoetida*, *Pseudomonas aeruginosa*, AMR (antimicrobial resistance), anti-pathogenic, *Caenorhabditis elegans*

INTRODUCTION

Ferula asafoetida L. is a herbaceous plant belonging to the family *Apiaceae*. This plant is widely distributed in arid zones of Iran and Afghanistan (Sonigra and Meena, 2021). A strong-smelled, tenacious and sulfurous oleogumresin obtained from this plant is considered to be of medicinal importance (Sepahi et al., 2015). In Hindi, it is called 'Hing', and as a spice used in virtually every kitchen across India. It is used to add taste and fragrance to many Indian dishes. Though India accounts for almost 40% of the world's total asafoetida consumption, in absence of any notable cultivation within the country, it satisfies almost all of its asafoetida demand by importing this resin from Afghanistan, Iran, and Uzbekistan [http://www.nationalfoods.co.in/]. Asafoetida has been part of folk medicine for centuries in India and some middle-east countries. Traditional Indian medicine of *Ayurved* prescribes asafoetida as an anti-flatulence agent (Amalraj and Gopi, 2017). This plant is also used to alleviate bacterial infections in western Iran. Though the essential oil from this plant has been reported to possess a wide array of bioactivities (Table 1), majority of such investigations are preliminary *in vitro* laboratory reports. For bringing the full benefits of the bioactive fractions/

components of asafoetida to the public, systematic and rigorous *in vivo* assays followed by clinical studies are necessary. In this chapter, besides reviewing the reports describing antimicrobial properties of asafoetida, we also report our own work on antibacterial activity in essential oils sourced from different asafoetida varieties.

Table 1. Various biological activities reported in *Ferula asafoetida*

No.	Type of activity reported	Remarks	References
1	Antidiabetic	*Ferula asafoetida* oleo-gum-resin ethanolic extract at 150 mg/kg body weight exerted significant antidiabetic activity in Wistar rat model.	Latifi et al., 2019.
2	Antihyperlipidemic	Treatment with *F. asafoetida* oleo-gum-resin extract decreased the serum levels of Total cholesterol (TC) and low density lipoprotein (LDL)-cholesterol; Serum level of high density lipoprotein (HDL)-cholesterol was significantly increased in comparison to diabetic control group of Wistar rats.	
3	Hemato-immunological parameters and growth performance	Dietary ferula powder (20-25 g kg^{-1}) seemed an effective herbal dietary supplement for enhancing humoral innate immune responses and growth of koi carp.	Safari et al., 2019
4	Mucosal antibacterial activity	The skin mucus of fish fed with Ferula powder-supplemented diets (20 g kg^{-1}) displayed antibacterial effect by inhibiting the growth of *Streptococcus iniae, Aeromonas hydrophila, Micrococcus luteus, Streptococcus faecium, and Yersinia ruckeri.*	
5	Antiproliferative effect	The ethanolic extract of *F. asafoetida* (2.5-10 μM) prevented the growth of PC12 and MCF7 cancer cells by inducing apoptosis, and caused a reduction in cell viability, as measured in MTT assay, in a time- and dose-dependent fashion.	Abroudi et al., 2020

Table 1. (Continued)

No.	Type of activity reported	Remarks	References
6	Anticancer activity	Essential oil of the seed of *F. asafoetida* (0.01-10 μl/mL), displayed anticancer activity in adenocarcinoma gastric (AGS) cell line.	Bagheri and Shahmohamadi, 2020
7	Neuroprotective effect	Asafoetida exhibited a neuroprotective role in cuprizone (CPZ)-induced mice models of multiple sclerosis by reducing death of neuronaldemyelination and oligodendrocytes at 25 and 50 mg/kg concentrations.	Bagheri et al., 2020
8	Antiproliferative effect	Dithiolane rich *F. asafoetida* EO displayed antiproliferative activity in human liver carcinoma cell lines (HepG2 and SK-Hep1) in a dose-dependent manner. Observed effects included induction of apoptosis and altered NF-κB and TGF-β signaling along with increased caspase-3 and TNF-α expression.	Verma et al., 2019

A Brief Review of Antimicrobial Activity of *F. asafoetida*

Quite a few published reports describe antimicrobial property of *F. asafoetida* (Table 2). However many such reports are limited to a semi-quantitative disc-diffusion assay, and do not report MIC (Minimum Inhibitory Concentration) values. Much work remains to be done with respect to confirmation and validation of such reported antimicrobial activities in different parts of this plant by different laboratories. Isolation of active fractions/compounds responsible for the antimicrobial activity is also warranted.

Table 2. Antimicrobial activity in *F. asafoetida*

No.	Remarks	References
1	Ethyl acetate, ethanol, and methanol extracts displayed antimicrobial activity against *Bacillus subtilis*, *Staphylococcus aureus, Klebsiella pneumonia, Escherichia coli, Aspergillus niger, Candida albicans,* with methanolic extract being the most potent. The MIC of these extracts against most of the test microorganisms ranged from 1-2 mg/mL respectively.	Patil et al., 2015
2	The essential oils inhibited growth along with conidia germination and germ tube elongation in *Penicillium digitatum* and *P. italicum.*	Zahani and Kahledi, 2018
3	Hexane extract was found to be effective against *Shigella flexineri*, but *S. aureus* was not affected much.	Bhatnager and Dang, 2015
4	The antibacterial activities of the EO were tested against standard strains of *Escherichia coli* (ATCC 25922), *Staphylococcus aureus* (ATCC 29312, 25923 and 700698), *Streptococcus mutans* (ATCC 35668), *S. sanguis* (ATCC 10556), *Enterococcus faecalis* (ATCC11700), *Bacillus cereus* (ATCC 11778), *Pseudomonas aeruginosa* (ATCC 27853), and clinical isolates of *Listeria monocytogenes*, *S. aureus, S. mutans, P. aeruginosa* and *E. coli*. The EO was active against all gram-positive and gram-negative bacteria at the concentration range of 0.5-8 μL/mL and 16-128 μL/mL, respectively. The EO inhibited the growth and killed all tested yeasts at 0.03-1 μL/mL. Moreover, the EO exhibited antifungal activity against the clinical and standard strains of filamentous fungi at 0.06-2 μL/mL.	Zomorodian et al., 2018
5	Pathani variety of asafoetida was effective against *Escherichia coli* and *Bacillus subtilis* with 18 mm and 28 mm zone of inhibition, respectively. Its MIC against both these bacteria was ~5 ppm. Volatile oil of *Irani* variety could inhibit growth of *Penicillium chrysogenum* and *Aspergillus ochraceus* by 70-75%, while that of Pathani variety inhibited fungal growth by 45-49%.	Divya et al, 2014
6	MIC of leaf extracts against *Escherichia coli, Staphylococcus aureus, Aspergillus niger,* and *Saccharomyces cerevisiae* were 62.5-125 mg/l, while it ranged from 50-400 mg/l in case of gum extracts.	Niazmand and Razavizadeh, 2020

Table 2. (Continued)

No.	Remarks	References
7	The essential oil from *F. asafoetida* oleo-gum-resin displayed notable antibacterial property against *Streptococcus mutans*, *Streptococcus sobrinus, Streptococcus sanguis, Streptococcus salivarius,* and *Lactobacillus rhamnosus.*	Daneshkazemi et al., 2019
8	The asafoetida gum silver nanoparticles (As-AgNPs) exhibited significant antimicrobial activity towards *E. coli, K. pneumoniae* and *C. albicans.* The MIC of the synthesized As-AgNPs was 7.80 μg/mL against *E. coli* and *S. typhi*, 31.20 μg/mL against *K. pneumonia* and *S. aureus*, and 15.60 μg/mL against *S. typhimurium*. The As-AgNPs registered a MIC of 15.60 μg/mL against the yeast *C. albicans.*	Devanesan et al., 2020
9	Essential oil (0.25%) extracted from resin could inhibit *Alternaria alternata, A. solani, A. flavus, A. niger, A. wentii, Rhizoctoniaspp. Drechsleratetramera, D. hawaiiensis, Fusarium semitectum, F. moniliforme* and *F. solani* isolated from okra seeds.	Sitara et al., 2018
10	Essential oil (2-3% v/v) extracted from latex reduced the production of violacein and pyocyanin pigments in *C. violaceum* and *P. aeruginosa*, respectively.	Khambhala et al., 2016
11	Essential oil (4 μL/mL) completely inhibited biofilm formation of *Candida albicans, C. dubliniensis,* and *C. krusei.*	Zomordian et al., 2018
12	Essential oil caused reduction in pyocyanin, pyoverdine, elastase, biofilm and homoserin lactones (HSL) production in *P. aeruginosa*. It also inhibited violacein production in *C. violaceum* extracted from foliar parts of *F. asafoetida.*	Sepahi et al., 2015

Our Work on Antimicrobial Potential of *F. Asafoetida* Essential Oil

Plant Material and Extraction

Resin of *F. asafoetida* was procured from National Foods (Baroda, India). Four different high-altitude varieties of asafoetida used were *Kabuli*, *Uzbeki, Shiro*, and *Irani*. Taxonomic identity of the samples was authenticated by Dr. Vasant Patel, Department of Botany, Smt. S. M. P. Science College, Gujarat, India. For extracting the essential oil (EO), the oleo gum resin was subjected to microwave (750W)-assisted hydrodistillation at 100°C for 30 min. Extraction efficiency obtained was 8.3 ± 0.2%. Oil thus obtained was stored under refrigeration to minimize evaporation and any possible deterioration.

Test Organisms

Bacteria

Bacterial strains employed in this work are listed in Table 3.

Among the bacterial strains used in this study, *C. violaceum* and *P. aeruginosa* displayed multi-drug resistant phenotype.

Nematode Worm

Caenorhabditis elegans (N2 Bristol) was used as the model host for pathogenic bacteria. This worm was maintained on Nematode Growing Media (NGM; 3 g/L NaCl, 2.5 g/L peptone, 1 M CaCl2, 1 M MgSO4, 5 mg/mL cholesterol, 1 M phosphate buffer of pH 6, 17 g/L agar-agar) with *Escherichia coli* OP50 as the feed. *E. coli* OP50 was procured from LabTIE B. V., JR Rosmalen, the Netherlands. Worm population to be used for the *in vivo* assay was kept on NGM plates not seeded with *E. coli* OP50 for three days prior to assay.

Antimicrobial Assay

Different *in vitro* assays for quantifying bacterial growth and pigment formation were performed using protocols described in our earlier publications (Joshi and Patel, 2019; Patel et al., 2020).

Table 3. Bacterial strains used in this study

No.	Name	Source	Remarks
1	*Chromobacterium violaceum*	MTCC 2656	This gram-negative bacterium is widely employed as a model for studying bacterial quorum sensing (QS), as the violet pigment produced by this bacterium is regulated by QS. This is being recognized as a potential emerging pathogen (Kothari et al., 2017) with low incidence but high mortality.
2	*Shigella flexneri*	MTCC 1457	An important pathogen of gastrointestinal (GI) tract
3	*Staphylococcus hominis*	MTCC 4435	Commensal of human skin, known for producing compounds contributing to body odour
4	*Pseudomonas aeruginosa*	MTCC 3541	Notorious pathogen capable of causing infection at multiple body sites
5	*Vibrio cholerae*	MTCC 3906	An important GI-tract pathogen
6	*Staphylococcus epidermis*	MTCC 435	Part of normal skin flora of humans
7	*Staphylococcus aureus*	MTCC 737	A versatile pathogen capable of causing infection at multiple body sites
8	*Salmonella typhi*	MTCC 3231	An important GI-tract pathogen

MTCC: Microbial Type Culture Collection, Chandigarh, India.

In Vivo Assay

Test bacterium was incubated with the sub-MIC level of EO for 24 h. Following incubation, 0.1 mL of this bacterial suspension was mixed with 0.9 mL of the M9 buffer containing 10 worms (L3-L4 stage). This experiment was performed in 24-well (sterile, non-treated) polystyrene plates (HiMedia), and incubation was done at 22°C. Number of live vs. dead worms was recorded till 5 days by putting the plate (with lid) under light microscope (4X).

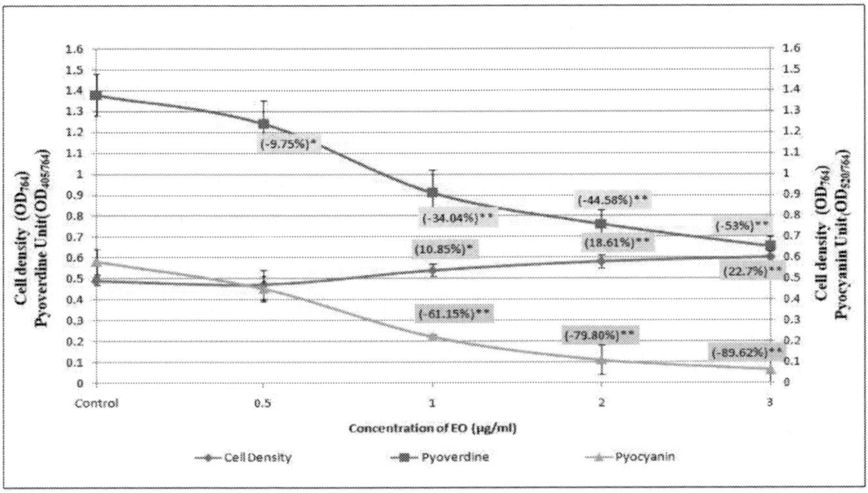

Figure 1. EO from Kabuli variety of asafoetida exerts inhibitory effect on QS-regulated pigment production in *P. aeruginosa*. $^{*}p < 0.05$; $^{**}p < 0.01$.

Results and Discussion

Kabuli Variety

Essential oil from the Kabuli variety of asafoetida was tested against *P. aeruginosa* and *C. violaceum*. This oil at 0.5-3%v/v was able to inhibit *P. aeruginosa*'s QS-regulated pigment production (Figure 1) and biofilm formation. Since exogenous supply of the QS-signal AHL (acyl homoserine lactone) did not reverse the inhibitory effect of essential oil on pigment production, this oil can be said to be acting as a 'signal-response inhibitor,' i.e., it may not be allowing the luxR component of the bacterial QS machinery to be functional. Besides acting as QS-inhibitor, this EO was also able to reduce the biofilm viability, and it could also eradicate the pre-formed biofilm of this notorious pathogen. Under the influence of this EO, *P. aeruginosa* also experienced a marginal reduction in its haemolytic and catalase activities. *P. aeruginosa* upon incubation with EO experienced a reduction in its ability to kill the model host worm *C. elegans* (Figure 2). Repeated subculturing of *P. aeruginosa* in media containing asafoetida EO was not found to induce any resistance. This is more likely to happen owing to more than one active ingredients in EO being responsible for the antibacterial activity, making the whole oil exert 'multiplicity of targets' against the susceptible pathogen.

The QS-inhibitory effect of EO was also evident in case of another gram-negative bacterium *C. violaceum*, wherein at 1%v/v EO inhibited violacein production by~75% (Figure 3). Growth of *V. cholerae* was inhibited in presence of this EO (25-150 μg/mL) by 60-70%.

Irani Variety

Oil from the Irani variety of asafoetida (25 μg-4 mg) was able to inhibit growth of *S. flexneri, S. typhi, S. epidermidis, S. hominis,* and *V. cholerae* (Table 5).

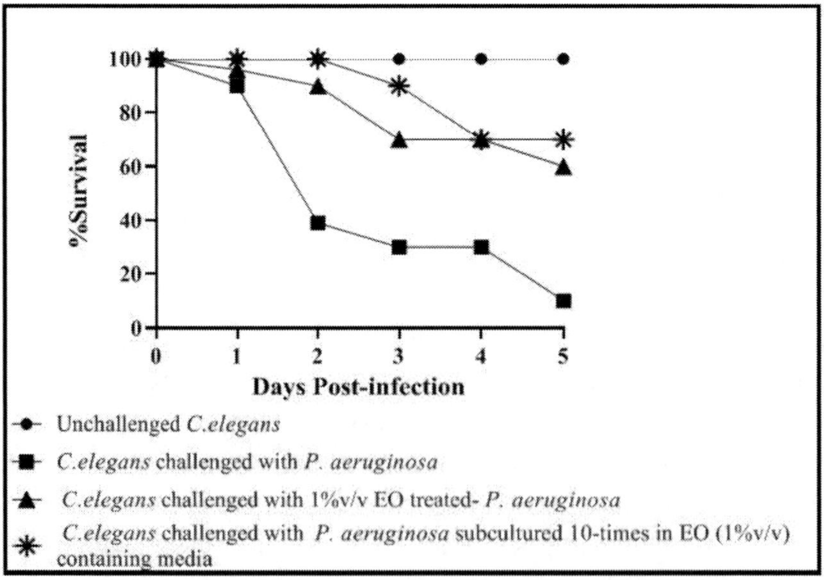

Figure 2. EO from Kabuli variety of asafoetida reduced *P. aeruginosa*'s virulence towards the *C. elegans* host.

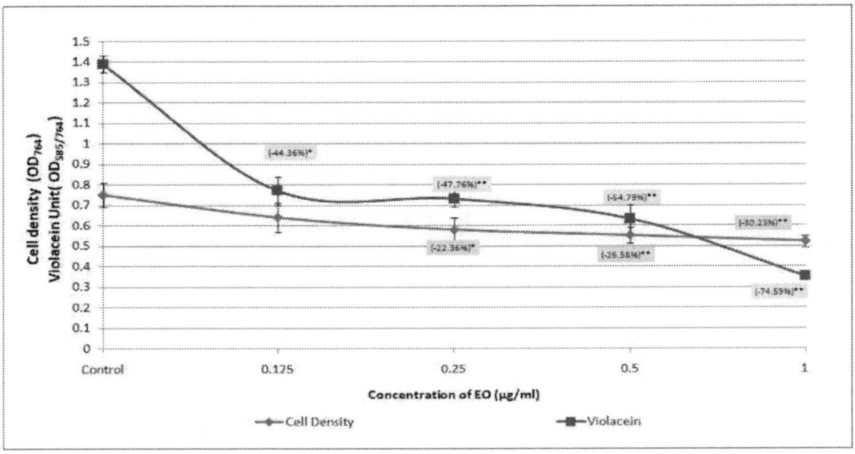

Figure 3. EO from Kabuli variety of asafoetida exerted inhibitory effect on QS-regulated pigment production in *C. violaceum.*

Table 4. Phytocompounds reported in *F. asafoetida* essential oil

Phytocompounds detected in the EO from Kabuli variety of asafoetida used in this study (A)	Phytocompounds reported as constituents of *F. asafoetida* essential oil by different labs (B)	Reference for column (B)
• 1,2-dithiolane • n-propyl sec-butyl disulfide • sec-butanethiol propyl disulfide • sec-butyl disulfide • 3,6-dimethyl-1,4-dithiane-2,5-dione • 1,3-dithiane • Allyl sulfone	• α-pinene • (Z)-propenyl sec-butyl disulfide • β-pinene • (E)-1-propenyl sec-butyl disulfide • (Z)-β-ocimene • (E)-β-ocimene	Estekhdami and Dehsorkhi (2008); Karimian et al., (2019)

Uzbeki Variety

Oil from the Uzbeki variety of asafoetida was able to inhibit growth of *S. flexneri, S. typhi, S. epidermidis, S. hominis,* and *V. cholerae*. It could also reduce the virulence of *P. aeruginosa* PAO1 against the nematode host *C. elegans* (Figure 4).

Table 5. Growth-inhibitory activity of *F. asafoetida* essential oils against different bacteria

Plant variety	Conc (mg/mL)	*Shigella flexeneri*	*Salmonella typhi*	*Staphylococcus epidermis*	*Staphylococcus hominis*	*Vibrio cholerae*
		% inhibition of growth				
Irani	0.025	16	60	35	1	5
	0.05	15	60	35	3	4
	0.075	-	60	50	2	-
	0.1	16	60	45	5	15
	0.125	16	65	55	4	25
	0.15	16	55	60	5	35
	0.2	-	60	-	60	-
	0.25	-	60	-	-	45
	0.3	-	60	-	62	-
	0.35	-	60	-	-	-
	0.4	25	60	-	65	-
	0.5	30	55	-	70	50
	1	55	50	-	70	50
	2	50	-	-	-	50
	3	50	-	-	-	50
	4	50	-	-	-	60
Uzbeki	0.025	18	62	50	3	3
	0.05	18	61	50	3	2
	0.075	19	62	55	3	-
	0.1	20	61	50	5	5
	0.125	20	61	60	5	10
	0.15	20	61	60	5	40
	0.2	20	62	-	70	-
	0.25	-	62	-	-	75
	0.3	20	61	-	70	-

Plant variety	Conc (mg/mL)	*Shigella flexeneri*	*Salmonella typhi*	*Staphylococcus epidermis*	*Staphylococcus hominis*	*Vibrio cholerae*
Uzbeki	0.35	-	60	-	-	-
	0.4	20	60	-	75	-
	0.5	25	61	-	75	80
	1	55	60	-	80	85
	2	55	-	-	-	85
	3	55	-	-	-	85
	4	55	-	-	-	85
Shiro	0.025	5	60	55	2	5
	0.05	5	60	60	3	4
	0.075	5	60	65	2	-
	0.1	5	60	70	3	10
	0.125	10	60	60	2	20
	0.15	15	60	70	5	25
	0.2	20	60	-	5	-
	0.25	-	60	-	-	20
	0.3	25	60	-	4	-
	0.35	-	60	-	-	-
	0.4	25	60	-	10	-
	0.5	30	60	-	15	25
	1	60	50	-	35	55
	2	65	-	-	-	80
	3	70	-	-	-	80
	4	85	-	-	-	85

EO from Kabuli variety of asafoetida was able to inhibit *V. cholerae*'s growth by 60-70% at 0.025-0.15 mg/mL.

Pre-treatment of *P. aeruginosa* PAO1 with this oil enhanced this pathogen's susceptibility to ampicillin by around 10% (data not shown).

Shiro Variety

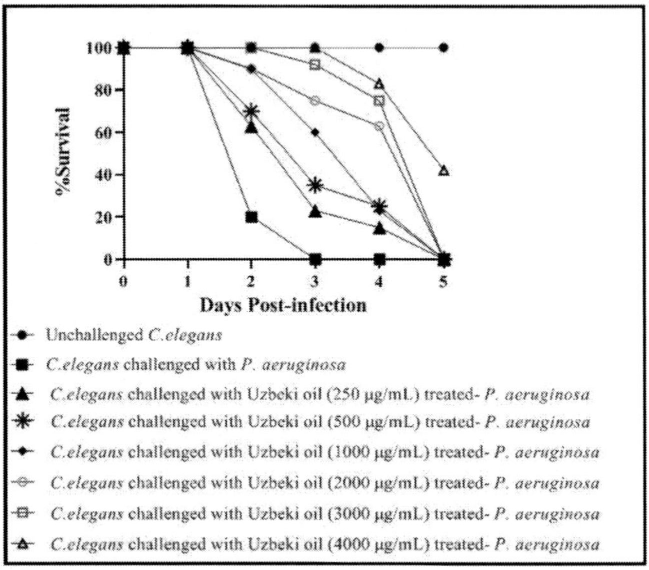

Figure 4. EO from Uzbeki variety of asafoetida reduced *P. aeruginosa*'s virulence towards the *C. elegans* host.

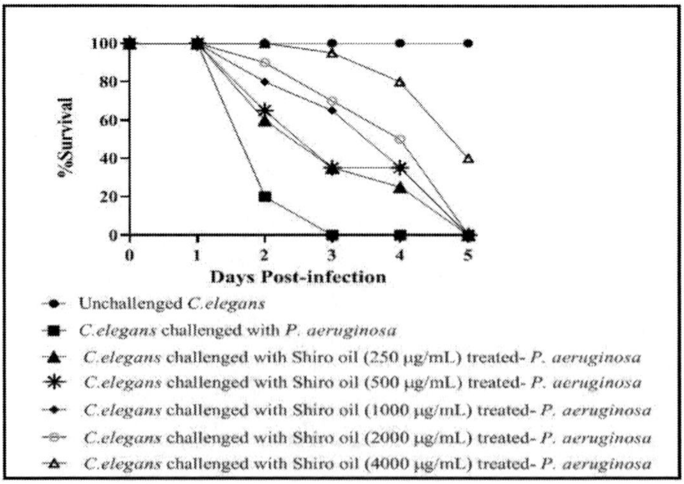

Figure 5. EO from Shiro variety of asafoetida reduced *P. aeruginosa*'s virulence towards the *C. elegans* host.

Oil from this variety of asafoetida was able to inhibit growth of *S. flexneri, S. typhi,* and *S. epidermidis*. This oil also could compromise *P. aeruginosa* PAO1's ability to kill the model host *C. elegans* (Figure 5). Pre-treatment of *P. aeruginosa* PAO1 with this oil enhanced this pathogen's susceptibility to chloramphenicol by around 10%.

Chemical Composition

Chemical composition analysis through gas chromatography revealed presence of different phytocompounds in Kabuli variety of EO (Table 4) (Verma et al., 2019). For a comparative overview, we list here compounds of asafoetida oil reported by other labs.

Final Comments

From our and other researcher's work, it can be said that asafoetida essential oil does possess antimicrobial activity. However from most reports, the crude oil appears to be effective against pathogens only at relatively higher concentrations. It is therefore logical to think that oils from different varieties of this plant should be subjected to chromatographic fractionation and individual fractions thus obtained should be tested against susceptible pathogens for their possible anti-pathogenic activity. This may result in identification of potent lead compounds for novel antimicrobials, and further investigation on their mode of action at molecular level can also help identify new targets in pathogenic bacteria. Since *F. asafoetida* essential oil has been reported to exert inhibitory effect on growth of Streptococci and *S. hominis* type of organisms, which are associated with oral cavity infections and body odour respectively, this oil can be said to have potential scope for deodorant and dentifrice applications too.

REFERENCES

Abroudi, M., Fard, A. G., Dadashizadeh, G., Gholami, O., & Mahdian, D. (2020). Antiproliferative Effects of *Ferula assa-foetida*'s Extract on PC12 and MCF7 Cancer Cells. *International Journal of Biomedical Engineering and Clinical Science*, 6(3), 60.

Amalraj, A., & Gopi, S. (2017). Biological activities and medicinal properties of *Asafoetida*: A review. *Journal of Traditional and Complementary Medicine*, 7(3), 347-359.

Bagheri, S. M., & Shahmohamadi, A. (2020). Anticancer Effect of Essential Oil of Seed of *Ferula assa-foetida* on Adenocarcinoma Gastric Cell Line. *International Journal of Clinical and Experimental Physiology*, 7(3), 96-99.

Bagheri, S. M., Maghsoudi, M. J., & Yadegari, M. (2020). Preventive effect of *Ferula asafoetida* oleo gum resin on histopathology in cuprizone-induced demyelination mice. *International Journal of Preventive Medicine*, 11.

Bhatnager, R., Rani, R., & Dang, A. S. (2015). Antibacterial activity of Ferula asafoetida: a comparison of red and white type. *Journal of Applied Biology and Biotechnology*, 3, 18-21.

Daneshkazemi, A., Zandi, H., Davari, A., Vakili, M., Emtiazi, M., Lotfi, R., &Masoumi, S. M. R. (2019). Antimicrobial activity of the essential oil obtained from the seed and oleo-gum-resin of *Ferula assa-foetida* against oral pathogens. *Frontiers in Dentistry*, 16(2), 113.

Devanesan, S., Ponmurugan, K., AlSalhi, M. S., & Al-Dhabi, N. A. (2020). Cytotoxic and antimicrobial efficacy of silver nanoparticles synthesized using a traditional phytoproduct, asafoetida gum. *International Journal of Nanomedicine*, 15, 4351.

Divya, K., Ramalakshmi, K., Murthy, P. S., & Rao, L. J. M. (2014). Volatile oils from *Ferula asafoetida* varieties and their antimicrobial activity. *LWT-Food Science and Technology*, 59(2), 774-779.

Estekhdami, P., & Dehsorkhi, A. N (2019). Chemical Composition of Volatile Oil of *Ferula assafoetida* L. *International Journal of Reseach Studies in Agricultural Sciences*, 5(8), 9-14.

Joshi, C., Patel, P., & Kothari, V. (2019). Anti-infective potential of hydroalcoholic extract of *Punica granatum* peel against gram-negative bacterial pathogens. *F1000Research*, 8.

Karimian, V., Ramak, P., & Majnabadi, J. T. (2019). Chemical composition and biological effects of three different types (tear, paste, and mass) of bitter *Ferula assa-foetida* Linn. gum. *Natural Product Research*, 1-6.

Khambhala, P., Verma, S., Joshi, S., Seshadri, S., & Kothari, V. (2016). Inhibition of bacterial quorum-sensing by *Ferula asafoetida* essential oil. *Adv. Genet. Eng.*, 5(2), 2169-0111.

Kothari, V., Sharma, S., & Padia, D. (2017). Recent research advances on *Chromobacterium violaceum. Asian Pacific Journal of Tropical Medicine,* 10(8), 744-752.

Latifi, E., Mohammadpour, A. A., Fathi, B., &Nourani, H. (2019). Antidiabetic and antihyperlipidemic effects of ethanolic *Ferula assa-foetida* oleo-gum-resin extract in streptozotocin-induced diabetic wistar rats. *Biomedicine & Pharmacotherapy*, 110, 197-202.

Niazmand, R., & Razavizadeh, B. M. (2020). *Ferula asafoetida*: chemical composition, thermal behavior, antioxidant and antimicrobial activities of leaf and gum hydroalcoholic extracts. *Journal of Food Science and Technology*, 1-12.

Patel, P., Joshi, C., Palep, H., & Kothari, V. (2020). Anti-infective potential of a quorum modulatory polyherbal extract (*Panchvalkal*) against certain pathogenic bacteria. *Journal of Ayurveda and Integrative Medicine*, 11(3), 336-343.

Patil, S. D., Shinde, S., Kandpile, P., & Jain, A. S. (2015). Evaluation of antimicrobial activity of asafoetida. *International Journal of Pharmaceutical Sciences and Research*, 6(2), 722.

Safari, O., Sarkheil, M., & Paolucci, M. (2019). Dietary administration of ferula (*Ferula asafoetida*) powder as a feed additive in diet of koi carp, Cyprinuscarpio koi: effects on hemato-immunological parameters, mucosal antibacterial activity, digestive enzymes, and growth performance. *Fish Physiology and Biochemistry*, 45(4), 1277-1288.

Sepahi, E., Tarighi, S., Ahmadi, F. S., & Bagheri, A. (2015). Inhibition of quorum sensing in Pseudomonas aeruginosa by two herbal essential oils from Apiaceae family. *Journal of Microbiology*, 53(2), 176-180.

Sitara, U., Akbar, A., Abid, M., & Ahmad, A. (2018). Essential oils show antifungal activity against seed-borne mycoflora associated with okra seeds. *International Journal of Biotechnology and Molecular Biology Research,* 15, 855-863.

Sonigra, P., & Meena, M. (2021). Metabolic profile, bioactivities, and variations in the chemical constituents of essential oils of the Ferula genus (Apiaceae). *Frontiers in Pharmacology*, 11, 2328.

Verma, S., Khambhala, P., Joshi, S., Kothari, V., Patel, T., & Seshadri, S. (2019). Evaluating the role of dithiolane rich fraction of *Ferula asafoetida* (apiaceae) for its antiproliferative and apoptotic properties: in vitro studies. *Experimental Oncology*, 4141(22), 90-94.

Zahani, F. H., & Khaledi, N. (2018). Biological effects of various essential oils on citrus decay pathogens. *International Journal of New Technology and Research,* 4(4), 263069.

Zomorodian, K., Saharkhiz, J., Pakshir, K., Immeripour, Z., & Sadatsharifi, A. (2018). The composition, antibiofilm and antimicrobial activities of essential oil of *Ferula assa-foetida* oleo-gum-resin. *Biocatalysis and Agricultural Biotechnology,* 14, 300-304.

In: Pathogenic Bacteria
Editor: Keith D. Watts

ISBN: 978-1-68507-422-7

Chapter 5

SONIC STIMULATION AT CERTAIN FREQUENCIES CAN CONFER LIMITED PROTECTION ON NEMATODE HOST INFECTED WITH *SERRATIA MARCESCENS*

Gemini Gajera, Purvi Garsondiya, Anjali Kalla, Himani Zaveri, Pinal Sanandiya, Drashti Patel, Nidhi Patel, Megha Maheshwari and Vijay Kothari *

Institute of Science, Nirma University, Ahmedabad, India

ABSTRACT

Effect of audible sound waves on bacteria has remained a poorly investigated area. We have been investigating microbial response to sonic stimulation for the last several years. This chapter describes a part of our work in this field, wherein different mono-frequency and poly-frequency sounds were investigated for their possible therapeutic effect on infected worm (*Caenorhabditis elegans*) population. Sound corresponding to the frequency of 400 Hz, 700 Hz and 2,000 Hz were

* Corresponding Author's E-mail: vijay.kothari@nirmauni.ac.in; vijay23112004@yahoo.co.in.

found to confer 11-31% survival benefit on worm population challenged with multi-drug resistant gram-negative pathogen *Serratia marcescens*. A combination of sound pattern and the antibiotic chloramphenicol was found to be more effective at saving worms from bacterial attack than either sound or antibiotic alone. Further investigations to elucidate the mechanistic details are warranted.

Keywords: Sonic stimulation, *Caenorhabditis elegans*, *Serratia marcescens*, bacterial infection

Background

Antimicrobial resistance (AMR) is being globally recognized by the policy makers as an immediate challenge requiring aggressive and coordinated response at the international level (Bhatia et al., 2019). In face of this serious health challenge impacting the global economy and development, novel innovative approaches not limited to discovery of new antibiotics are urgently warranted. Among the possible novel approaches, one non-invasive strategy can be testing the therapeutic potential of electromagnetic radiation or sound waves (sonotherapy) against antibiotic-resistant pathogenic bacteria. In this study, we investigated effect of a variety of mono-frequency and poly-frequency sounds on the model nematode host *Caenorhabditis elegans* infected with pathogenic bacteria.

Methods

Organisms

Caenorhabditis elegans (N2 Bristol) was used as the model host for pathogenic bacteria. This worm was maintained on Nematode Growing Media (NGM) with *Escherichia coli* OP50 as the feed, as previously described by us (Joshi et al., 2019). *E. coli* OP50 was procured from LabTIE B. V., JR Rosmalen, the Netherlands.

Pathogenic bacteria *Chromobacterium violaceum* (MTCC 2656), *Serratia marcescens* (MTCC 97), *Staphylococcus aureus* (MTCC 737), and *Streptococcus pyogenes* (MTCC 1924) were procured from Microbial Type Culture Collection (MTCC), Chandigarh. *Pseudomonas aeruginosa* was taken from our internal lab collection. Of the pathogenic strains mentioned above, all the gram-negative bacteria were multi-drug resistant, and the *C. violaceum* and *S. marcescens* strains used were beta-lactamase producers. Their antibiogram, along with culture conditions and growth media has previously been reported by us (Patel et al., 2019a).

Sonic Frequencies

Following sounds were used in this study:

Poly-Frequency Sound

Spotted Babbler bird's voice: 2260-2712 Hz.

Koel bird's voice: 602-1592 Hz.

Gayatri Mantra (GM) and *Mahamrityunjay Mantra* (MM): These are two spiritual chantings from ancient Indian tradition. The source sound was however with background music. Frequency of these sound files ranged from 0-15,000 Hz in case of GM, while this range for MM was 0-16,000 Hz. Seven and five such regions were respectively identified in GM and MM files wherein intensity of frequencies in the range of 0-4,000 Hz were prominent.

Mono-Frequency Sound

300 Hz, 400 Hz, 700 Hz, 1000 Hz, 2000 Hz

Bird voices recorded at Dang Forest, Ahwa, Gujarat were obtained from Dr. Arun Dholakia (Zoology Department, PTS Science College, Surat). Frequency analysis was executed by performing a temporal frequency analysis (TFFT) using NCH WavePad v 7.13.

Sound beeps of each required mono-frequency was generated using NCH® tone generator. Each of the sound files played during the experiment was prepared using WavePad Sound Editor Masters Edition v 7.13 in such a way that there is a time gap of one second between two consecutive beeps of mono-frequency sounds (each beep was of 1 s duration). No time gap was there between two consecutive playing of bird voice. MM and GM files were sourced from a DVD (J. K. Movies, Kolkata) available in local market.

Files containing those sounds which yielded positive results in this study are available as supplementary material at: https://doi.org/10.31219/osf.io/kewps

Sound intensity was measured with digital sound level meter KM929MK-1 (Kusam-Meco Import Export Pvt. Ltd., Navi Mumbai). This sound level meter detects the sound intensity within the range of 35 dB to 130 dB.

Sonic Stimulation of the Infected Worms

Establishment of bacterial infection in the host worm, and sonic stimulation of the infected worms was executed as described in our previous study (Patel et al., 2019b). Spotted babbler's sound was played at the intensity of 64-78 dB, and that of Koel at 66-80 dB. Background noise was recorded to be 64 dB. Mono-frequency sounds were played at 85.5 dB. Sound intensity level for MM and GM patterns was 75.5-85.5 dB and 73.7-90.3 dB respectively. Effect of sound treatment on infected worms was quantified through a live-dead assay, wherein number of live vs. dead worms was counted on daily basis till five days by putting the assay plate (with lid) under light microscope (4X).

Statistical Analysis

All the experiments were performed in triplicates, and measurements are reported as mean ± standard deviation (SD) of three independent experiments.

Statistical significance of the data was evaluated by applying t-test using Microsoft Excel®. P values ≤0.05 were considered to be statistically significant.

RESULTS

When the worms infected with different pathogenic bacteria were subjected to any of the test sounds, no statistically significant difference in the number of surviving worms was observed between the control and experimental wells (data not shown), except the cases where *S. marcescens*-infected worms were subjected to (i) mono-frequency sound pertaining to 700 Hz (Figure 1A) or 2,000 Hz (Figure 1B) or 400 Hz (Figure 1C); and (ii) poly-frequency MM and GM patterns. Sonic stimulation of the infected worms with 400 Hz or 700 Hz sound seemed to be more effective than 2,000 Hz. 400 Hz and 700 Hz conferred 30.83% ($p < 0.001$) and 27% ($p < 0.05$) survival benefit respectively on the worm population facing *S. marcescens* challenge, as compared to 11% ($p < 0.001$) offered by latter, as revealed from worm count recorded on fifth day. Protective effect offered by 2000 Hz sound was marginally higher on fourth day than that on fifth day.

Since 400 Hz sound seemed to be the most effective against bacterial infection, we performed an additional experiment at this particular frequency. While in the experiment depicted in Figure 1, sonic stimulation was started soon after adding bacteria into the wells containing worms, in this additional experiment (Figure 2) we first allowed bacteria to establish infection for 6 h or 24 h, and then started the sonic stimulation. In this scenario too, wherein the pathogen is allowed sufficient time to establish the infection before the sonotherapy starts, sonic stimulation was able to support ~20% better survival of the worms in face of pathogen challenge.

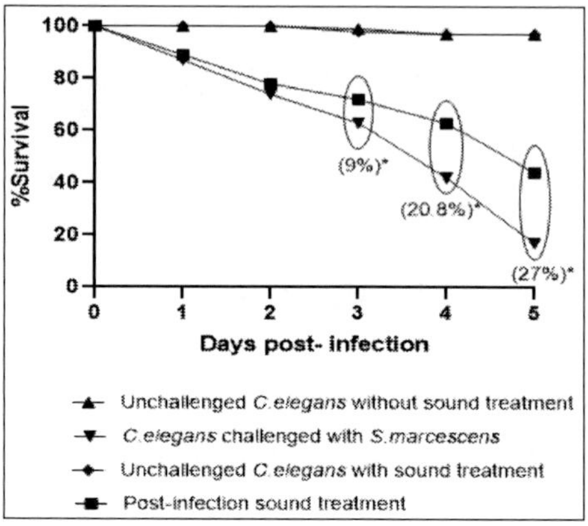

A

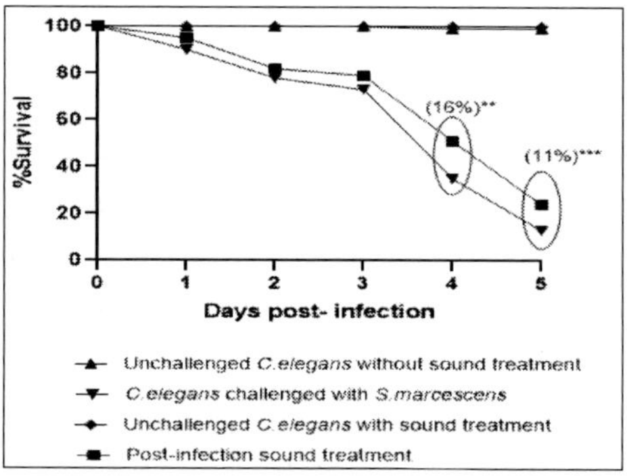

B

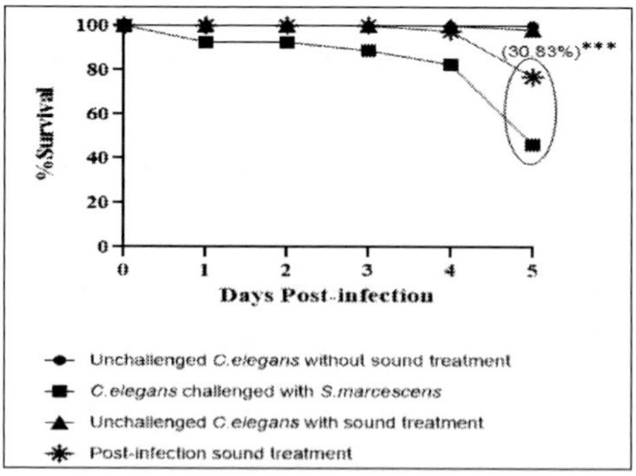

C

Figure 1. Protective effect of post-infection sound treatment on *C. elegans* challenged with *S. marcescens*. (A) (700 Hz; 85.5 dB); (B) (2000 Hz; 85.5 dB); (C) (400 Hz; 85.5 dB). Sound treatment had no effect on survival of non-infected worms; Percent values in parenthesis indicate the difference between number of surviving worm in control and experimental wells; $^{*}p < 0.05$, $^{**}p < 0.01$, $^{***}p < 0.001$.

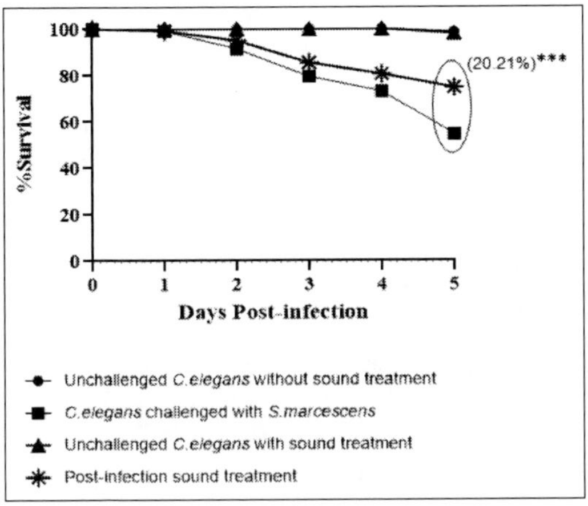

A

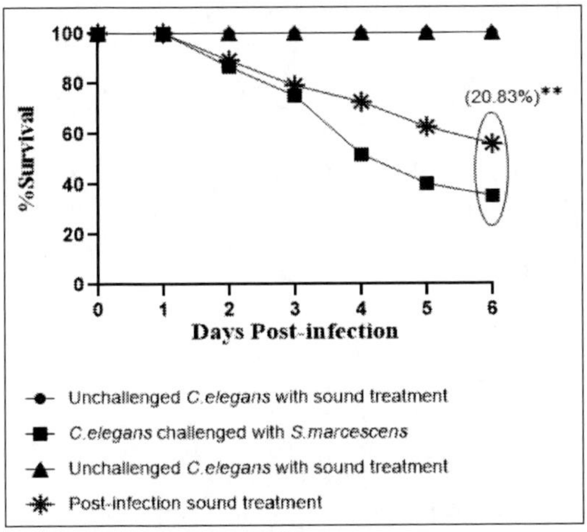

B

Figure 2. Protective effect of post-infection 400 Hz (85.5 dB) sound treatment on *C. elegans* challenged with *S. marcescens.* (A) Sonic stimulation started 6-h post-infection (B) sonic stimulation started 24-h post infection. Sound treatment had no effect on survival of non-infected worms; Percent values in parenthesis indicate the difference between number of surviving worm in control and experimental wells; $^{*}p < 0.05$, $^{**}p < 0.01$, $^{***}p < 0.001$.

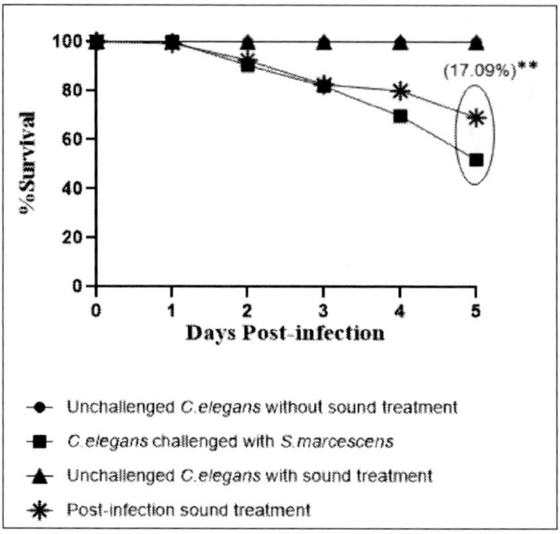

A

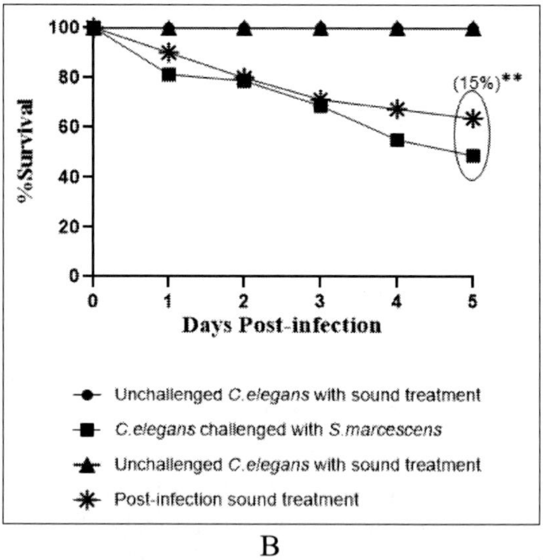

B

Figure 3. Effect of (A) *Gayatri Mantra* (73.7-90.3 dB); and (B) *Mahamrityunjay* Mantra (75.5-85.5) dB on infected worms. Sound treatment had no effect on survival of non-infected worms; Percent values in parenthesis indicate the difference between number of surviving worm in control and experimental wells; $^{**}p < 0.01$.

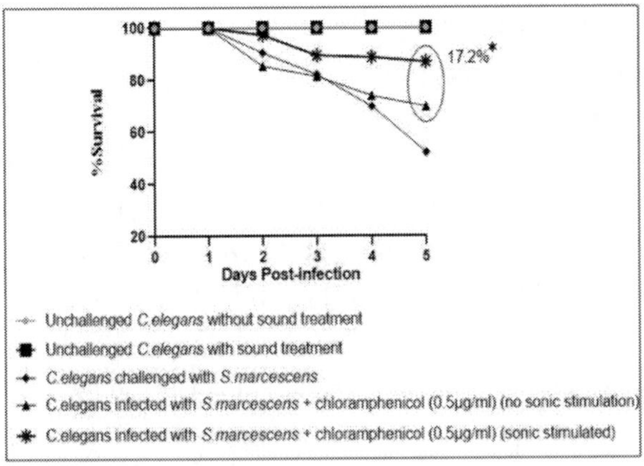

Figure 4. Combination of sonic stimulation and antibiotic [(*Gayatri Mantra*; 73.7-90.3 dB) + chloramphenicol (0.5 µg/mL)] offers better protection to the worms against bacterial challenge, than either sound or antibiotic alone. Sound treatment had no effect on survival of non-.infected worms; Percent values in parenthesis indicate the difference between number of surviving worm in control and experimental wells; $^{*}p < 0.05$.

Among the poly-frequency sound patterns, GM and MM were able to support marginally better survival of worm population challenged with *S. marcescens* (Figure 3). We also tried to see whether the combination of sonic stimulation and a bactericidal antibiotic (chloramphenicol; HiMedia) at sub-MIC level can be more effective than either sound or antibiotic (sub-MIC) alone. Such a sound + antibiotic combination was found to be effective more than either sound or antibiotic alone (Figure 4). While GM and antibiotic individually offered 17.80% and 16.96% protection to the worms against bacterial attack, their combination offered 34.18% protection.

CONCLUSION

Taken together with our previous results (Patel et al., 2019b), this study indicates that certain sonic frequencies (e.g., 400 Hz, 700 Hz, 2,000 Hz) or certain poly-frequency sound patterns can exert some therapeutic effect on worms challenged with pathogenic bacteria. This effect seem to be a function of the 'type' (e.g., frequency) of sound used, and also the pathogenic bacterial species being targeted. From the limited insight provided by our hitherto studies (Kothari et al., 2016; Kothari et al., 2018; Patel et al., 2019b) *S. marcescens* appear to be a sound-responsive bacterium. Since worms do not seem to be responsive to sound in these experiments, all the observed effect can be attributed to the interaction of sound with bacteria. Insensitivity of the wild type *C. elegans* to ultrasound has been reported by Ibsen et al. (2015) too. Challenge for future studies lies in elucidating the molecular mechanisms associated with this sound-responsiveness of bacteria, and identifying the most effective sonic stimulation (in terms of frequency and sound intensity) offering maximum protection against bacterial infections. Whether similar protective effect of sonic stimulation can be demonstrated in more complex model organisms, yet remains to be investigated. Though a long-long path remains to be travelled for developing an evidence-based sonotherapy, extensive research in this field over several years may offer some realistic probability of using sonic stimulation/ music for therapeutic purpose, i.e., musopathy (https://www.dtnext.in/News/City/2019/12/31013955/1206806/Musopathy-Mapping-music-and-its-medicinal-benefits.vpf).

Acknowledgments

Authors thank Nirma Education and Research Foundation (NERF, Ahmedabad) for infrastructural and financial support; and Dr. Arun Dholakia for providing bird voices recorded by him.

References

Bhatia, R., Katoch, V. M. and Inoue, H., 2019. Creating political commitment for antimicrobial resistance in developing countries.

Indian Journal of Medical Research, 149(2), p. 83. https://dx.doi.org/10.4103/ijmr.IJMR_1980_17.

Ibsen, S., Tong, A., Schutt, C., Esener, S., & Chalasani, S. H. (2015). Sonogenetics is a non-invasive approach to activating neurons in *Caenorhabditis elegans*. *Nature Communications*, 6(1), 1-12. https://doi.org/10.1038/ncomms9264.

Joshi, C., Patel, P. and Kothari, V., 2019. Anti-infective potential of hydroalcoholic extract of Punica granatum peel against gram-negative bacterial pathogens. *F1000Research*, 8. https://dx.doi.org/10.12688%2 Ff1000research.17430.2

Kothari, V., Joshi, C., Patel, P., Mehta, M., Dubey, S., Mishra, B. and Sarvaiya, N., 2016. Influence of a mono-frequency sound on bacteria can be a function of the sound-level. *Indian Journal of Science and Technology,* 2018, 11. http://dx.doi.org/10.17485/ijst%2F2018%2F v11 i4%2F111366.

Kothari, V., Patel, P., Joshi, C., Mishra, B., Dubey, S. and Mehta, M., 2016. Quorum sensing modulatory effect of sound stimulation on Serratia marcescens and Pseudomonas aeruginosa. *Current Trends in Biotechnology and Pharmacology,* 2016, 11, 121-128. https://doi.org/10.1101/072850.

Patel, P., Joshi, C. and Kothari, V., 2019a. Antipathogenic potential of a polyherbal wound-care formulation (Herboheal) against certain wound-infective gram-negative bacteria. *Advances in Pharmacological Sciences,* 2019, p. 1-17. https://doi.org/10.1155/2019/1739868.

Patel, P., Patel, H., Vekariya, D., Joshi, C., Patel, P., Muskal, S. and Kothari, V., 2019b. Sonic stimulation, and low power microwave radiation can modulate bacterial virulence towards *Caenorhabditis elegans*. *Anti-Infective Agents*, 17, p. 1-13. https://doi.org/10.2174/2211352516666181102150049.

INDEX

A

B

E

F

G

H

I

L

M

N

O

P

Q

R

S

T

U

V

W